DARWIN
Devolves

DARWIN *Devolves*

The New Science About DNA That Challenges Evolution

MICHAEL J. BEHE

HarperOne
An Imprint of HarperCollins*Publishers*

 For information, address HarperCollins Publishers, 195 Broadway, New York, NY 10007.

HarperCollins books may be purchased for educational, business, or sales promotional use. For information, please email the Special Markets Department at SPsales @harpercollins.com.

FIRST EDITION

Designed by SBI Book Arts, LLC

Library of Congress Cataloging-in-Publication Data

Names: Behe, Michael J., author.
Title: Darwin devolves : the new science about DNA that challenges evolution / Michael J. Behe.
Description: First edition. | New York, NY : HarperOne, 2019 | Includes bibliographical references.
Identifiers: LCCN 2018034062 (print) | LCCN 2018040030 (ebook) | ISBN 9780062842688 (e-book) | ISBN 9780062842619 (hardcover) | ISBN 9780062842664 (paperback) | ISBN 9780062842688 (digital edition)
Subjects: LCSH: Evolution (Biology) | DNA. | Molecular evolution. | Darwin, Charles, 1809–1882.
Classification: LCC QH367.3 (ebook) | LCC QH367.3 .B427 2019 (print) | DDC 576.8—dc23
LC record available at https://lccn.loc.gov/2018034062

19 20 21 22 23 LSC 10 9 8 7 6 5 4 3 2 1

To my friends and colleagues at the Discovery Institute—
especially Michael Denton and Phillip E. Johnson,
whose work profoundly influenced me

The First Rule of Adaptive Evolution:

Break or blunt any functional gene whose loss would increase the number of a species's offspring.

CONTENTS

Introduction

When I was a kid I would lie awake some nights pondering existential questions: What is thought? Why am I me? How did the world get here? I admit I was a peculiar boy, but over time I found that nearly all my friends had asked themselves such questions too. It seems to come naturally with having a mind. Most of the time we're distracted with everyday activities—TV, school, dinner. But once in a while, in a quiet moment, we realize that something completely different must have happened to give rise to what we call ordinary life.

Later I learned that not only young people ask that last question; young civilizations do too. Discussion of the enigma of where nature came from goes back as far as there are written historical records and, with a few lulls, has continued strongly up to the present. Yet despite the long and varied history of discourse, all *particular* positions on the topic can be considered to be elaborations on either of just two *general* mutually exclusive views: (1) contemporary nature, including people, is an accident; and (2) contemporary nature, especially people, is largely intended—the product of a preexisting reasoning mind.

I will argue in this book that recent progress in our understanding of the molecular foundation of life *decisively* supports the latter view. To help frame the issues we'll consider later, let's first briefly recall a few highlights of what earlier writers thought about nature and purpose.

Throughout History

The first person known to have discussed the likelihood of teleology—purpose—in nature was a Greek named Anaxagoras, who was born about the year 510 BCE in a region that's now part of Turkey.[1] He thought, roughly, that the elements of matter originally were chaotic, fragmented, and mixed, but were then purposely arranged into their present form by *nous*, the Greek term for "mind." His student Diogenes of Apollonia was even more explicit: "Without an intelligence it would not be possible that the substance of things should be so distributed as to keep all [nature] within due measure."

Now, remember, we're looking back on an era when the elements were thought to be earth, air, fire, and water; little was known then about the composition and properties of nature beyond what could be seen with the naked eye. What's more, the very ability to frame the right questions and deal with fair objections was still rudimentary. It turns out that the basic question nearly all reasoning people (even kids) ask, "Whence nature?" is much more involved than its length might suggest. Finding a good, justified answer necessarily depends on our understanding of both nature and logic. In turn, that means the answer depends on progress in both science and philosophy.

The epitome of science in the classical world was arguably the work of the second-century Roman physician Galen, who had a very definite point of view on the origin of nature. In his book *On the Usefulness of the Parts of the Body*, which provided a sophisticated functional analysis of its subject matter, Galen concluded that the human body is the result of a "supremely intelligent and powerful divine Craftsman," that is, "the result of intelligent design."[2]

Not everyone in ancient times, however, was on board with that claim. Rejoinders to design included types of arguments we still see today, such as that a good designer wouldn't allow humans to suffer and that no designer would make such foul creatures as moths and snakes. A contrarian school of thought called atomism held that nature was composed of just atoms and void and that occasionally by serendipity atoms would aggregate into something larger. Like a primitive form of Darwin's theory, the argument continued that if perchance the aggregate formed an organism that could survive, then it survived; if not, it didn't; so it's no surprise that we now see what we see, you see. Critics retorted that they never saw particles coming together by chance to form even a simple house, let alone an enormous complicated universe.

When Christianity appeared, the design view gained a new source of support. The second-century Christian writer Tertullian pointed to perceived workmanship in the forms and functions of insects. The contemporary theologian Origen argued that the skill needed to construct animals indicated the highest intelligence. The great fourth-century philosopher-theologian Augustine of Hippo shared such views and added his own points, including that: (1) we see only facets of the design mosaic, and so can't fairly judge the whole; (2) the structures of the smallest creatures are as wonderful as those of the largest; and (3) humans are more remarkable than other animals because they possess reason—mind itself.

Over the next thousand years the topic was put on the back burner, perhaps because, with the establishment of Christianity as the dominant religion of the West, the designedness of nature was a widely shared view rather than a matter for dispute. However, the accelerating progress of both science and philosophy from the Middle Ages onward rekindled discussions. The sixteenth-century English philosopher Francis Bacon urged science to rely on inductive reasoning in its work and to separate itself from philosophy. (The two had overlapped considerably until then. In fact, what we now call "science" was then called "natural philosophy.") The eighteenth-century Scottish philosopher David Hume attacked inductive reasoning in general and the design argument in particular. He argued that, in order to think that *our* world was designed, we would need to have much experience examining *other* worlds that had been designed. Since we have no such experience, he concluded, the design argument is not justified. Several decades later, the Anglican clergyman William Paley, ignoring Hume and drawing on sophisticated work in biology, presented the watchmaker argument (discussed in Chapter 3)—widely considered to be the strongest, most detailed case for design up until his day.

About sixty years later Charles Darwin parried Paley's argument. He proposed that there was a hitherto unrecognized natural process that, over a very long time, could imitate the results of purposeful design—namely, natural selection acting on random variation. That contention obligated design proponents to dispute its plausibility at an intricate biological level, so the depth and breadth of knowledge required for meaningful discussion skyrocketed. In practice, although most biologists of his day were skeptical of Darwin's proposed mechanism of evolution, the very broaching of a seemingly plausible nondesign explanation led most scientists to abandon the idea of a discernible purpose in the structures of life, so few were left to argue the point.

Recall, however, that the state of the design argument depends on our understanding of science and logic, which has accelerated explosively since Darwin's day. The development of analytical philosophy in the early twentieth century encouraged much more rigorous arguments; advances in formal logic and probability theory, such as Bayes' theorem, made that easier.[3] What's more, not all scientists had abandoned design. Among them was Alfred Russel Wallace, who, along with Darwin, is credited with being the cofounder of the theory of evolution.

Wallace thought that much of nature showed strong evidence of purpose, as he forcefully conveyed in *The World of Life: A Manifestation of Creative Power, Directive Mind and Ultimate Purpose.*[4] In other words, in modern parlance, *the very cofounder of the theory of evolution was an intelligent-design proponent.* In 1910 the chemist Lawrence Henderson first noticed that the environment of the earth was remarkably fit for life,[5] and, despite naive early ideas about the likelihood of life on Mars and elsewhere, exploration showed space to be desolate. Subsequent progress concluded that it's not just our world—the physics and chemistry of the whole universe is astonishingly fine-tuned for intelligent life on earth.[6] And, of course, as I'll emphasize in this book, in the late twentieth and early twenty-first centuries biology unexpectedly discovered astounding sophisticated machinery at the molecular foundation of life.

I will contend that, for any who agree that they themselves have a mind (no, not everyone agrees, as we'll see in the final chapter) and whose mind is open on the question, those twentieth-century advances—together with even more crucial twenty-first-century ones that we'll explore—should *definitively* settle the broad basic issue in favor of design. Additional details of particularized claims, of course, remain open for lively disputation.

A Winding Road

But first a necessary digression to explain how I came to disagree with most contemporary scientists on this pivotal subject. Imagine my surprise a while back when I opened an academic journal called *Biology & Philosophy* and spotted this sentence: "To see the point quite palpably, note that Stalin, or Osama bin Laden, or Michael Behe, or your favorite villain is also . . ."[7] The man who included me in that rogues gallery was Alexander Rosenberg, R. Taylor Cole Professor of Philosophy at Duke University—a fellow I've never met.

His article had precious little to do with me. The line was an offhand remark in the course of arguing that the well-known philosopher Daniel Dennett—a founding member of the New Atheists—was something of a wimp, because in his books he didn't clearly spell out the utter nihilism that Rosenberg saw as a consequence of Darwin's theory.

It was a silly remark but, unfortunately, it does accurately reflect the hostility felt by a large chunk of academia toward those of us who publicly argue the case for purpose in nature. (Notice that the overt insult was passed along by the reviewers of the article and the journal editor.) We might see ourselves as just trying to puzzle out those existential questions that kept us awake at night as kids. But folks such as Rosenberg seem to envision peasants with torches and pitchforks marching on their faculty offices. We might just be wondering what the evidence of nature really shows. But "since nihilism is true,"[8] too many academics think there's nothing to think about; therefore contrary views must be dishonest. So before we begin the book I want to try to head off such charges of bad faith. To show that I come by my views honestly, let me very briefly recount the history of my own thinking.

I was born into a large Roman Catholic family and, like all of my brothers and sisters, attended Catholic grade school and high school. Unlike some Christian denominations, the Catholic Church never had much of a problem with evolution. I remember being taught about it in seventh grade by Sister David Marie. The important point, she stressed, is that God created the universe, life, and humanity. How he did that, whether quickly or slowly, employing natural law or not, was up to him, not us, and our best evidence these days shows that evolution is correct. That view was perfectly fine with me. In fact, although I wasn't aware of it then, it had been the predominant understanding in Catholic circles for a long time. For example, the 1909 *Catholic Encyclopedia* has a lengthy scholarly article on evolution that makes a number of crucial distinctions, including a distinction "between the [basic] theory of evolution and Darwinism."[9] Plain "evolution" was no big theological deal. But framing it as necessarily nihilistic, as Alexander Rosenberg and many others do, was tantamount to denying Christianity. Even as a boy I had plenty of reasons to believe in God that had nothing to do with evolution.

When I went off to Drexel Institute of Technology (now Drexel University), I decided to major in chemistry, specifically because I wanted to know how the world *worked*; I wanted to know what made things tick. Since everything is made of chemicals, then chemistry seemed to be the obvious choice. During my college years I had a summer "co-op" job in a biochemistry lab at the Department of Agriculture research facilities near Philadelphia, where I became fascinated with the chemistry of life. Senior year at Drexel I took a course on evolutionary biochemistry to learn how it all came together.

During graduate studies in biochemistry at the University of Pennsylvania and postdoctoral work at the National Institutes of Health, I had no qualms about standard evolutionary theory and would occasionally (and smugly) tease friends who did. I remember one day at the NIH chewing over the Big Questions with a fellow Catholic postdoc, Joanne (her brother was a priest), who was in the same lab I was. Talk turned to the origin of life. Although she and I were both happy to think life started by natural laws, we kept bumping up against problems. I pointed out that to get the first cell, you'd first need a membrane. "And proteins," she added. "And metabolism," said I. "And a genetic code," said she. After a short time we both looked wide-eyed at each other and simultaneously shouted, *"Naaaahh!"* Then we laughed and went back to work, as if it didn't really matter to our views. I suppose we both thought that, even if *we* didn't know how undirected nature could begin life, *somebody* must know. That's the impressive power of groupthink.

After three years at my first job as an assistant professor at Queens College in New York City, my new wife, Celeste, our firstborn daughter, Grace, and I moved to Bethlehem, Pennsylvania, where a new job awaited at Lehigh University. Several very busy years later I paused to read a book that startled me and changed my view of evolution. *Evolution: A Theory in Crisis* by Michael Denton, a geneticist and medical doctor then teaching in Australia, offered no solution to the riddle of life, but pointed out numerous serious problems for Darwin's theory at the molecular level that I had never even heard about—even though I was a biochemistry professor whose goal in entering science was to understand how the world worked! At that point, when I thought back, I realized I had never heard any of my teachers critique Darwin's theory in all of my science studies.

I got mad. Over the following months I spent much time in the science library trying to find papers or books that explained in real detail how random mutation and selection could produce the exceedingly intricate systems routinely studied by biochemistry. I came up completely empty. Although many publications would pay homage to Darwin and a few would spin "Just So" evolutionary tales, none spelled out how his mechanism accounted for complex functional systems. Vague stories had kept me satisfied in the past, but no longer. Now I wanted real answers.

At that point I concluded that I had been led to believe in Darwin's theory not because of strong evidence for it. Rather, it was for sociological reasons—that simply was the way educated people were expected to think these days. My professors hadn't been intentionally misleading—that was the framework in which they thought about life too. But from then on I resolved to decide for myself what the evidence showed.

When one starts to treat Darwinism as a hypothesis about the biochemical level of life rather than as an assumption, it takes about ten minutes to conclude it's radically inadequate. It takes perhaps another ten minutes to realize that the molecular foundation of life was designed, and for effectively the same reason that Anaxagoras, Galen, and Paley reached the same conclusion for visible levels of biology (although, because of progress in science and philosophy, the argument is now necessarily much more detailed and nuanced than their versions): the signature of intelligent activity is the arrangement of disparate parts to fulfill some purpose. The molecular parts of the cell are elegantly arranged to fulfill many subsidiary purposes that must blend together in service of the large overall purpose of forming life. As we'll see in this book, *no* unintelligent, undirected process—neither Darwin's mechanism nor any other—can account for that.

With the aid of the then newfangled internet, over the years I met other academics who had had experiences roughly similar to mine, who had been perfectly willing to accept Darwinian evolution, but at some point realized with shock that the larger theory was an intellectual facade. Like me, most had religious convictions, which freed them from the crippling assumption that—no matter what the evidence showed—unintelligent forces simply *must* be responsible for the elegance of life. Some of us banded together under the auspices of the Seattle-based think tank Discovery Institute, the better to defend and advance the topic of intelligent design (ID), to which we had become dedicated.

In conversations with them I discovered that, as a biochemist, I had ideas to contribute that the others did not. At the urging of Phillip Johnson, then a professor of law at the University of California–Berkeley, I set about writing a book that in 1996 became *Darwin's Black Box: The Biochemical Challenge to Evolution*. Except for answering extravagant Darwinian claims or attacks on ID,[10] I thought I was done with writing at that point. But the rapid progress of science in the subsequent decade allowed further arguments to be made. In 2007 those became *The Edge of Evolution: The Search for the Limits of Darwinism*, which, as the title suggests, tried to locate the point in life where what can be explained solely by unintelligent forces is reached. (One common confusion of critics is to think that ID argues *everything* is planned. That's not the case. Chance is an important, if superficial, feature of biology.) Again I thought I was done, but even greater unanticipated progress in biology over the past ten years has spurred me to write this book.

Where We're Headed

The firm conclusion I've drawn over the past decades is this: despite occasional questions and bumps along the road, the greater the progress of science, the more deeply into life design can be seen to extend. In Darwin's own day, the mid-nineteenth century, scientists wondered whether there was sufficient variety in nature's creatures to fuel his theory. After DNA and proteins were discovered in the late twentieth century, a pressing question was whether Darwin's mechanism—natural selection acting on random mutation—could account for even the biochemical level of life and the sophisticated molecular machinery unexpectedly discovered there.

As science rapidly advanced in the early twenty-first century, large studies showed only surprisingly minor changes in genes under severe selective pressure. And as we'll see in this book, now several decades into the twenty-first century, ever more sophisticated studies demonstrate that, ironically, random mutation and natural selection are in fact fiercely *de*volutionary. It turns out that mutation easily *breaks* or *degrades* genes, which, counterintuitively, can sometimes *help* an organism to survive, so the damaged genes are hastily spread by natural selection. Strangely, in the space of a century and a half Darwinism has gone from the chief candidate for the explanation of life to a known threat to life's long-term integrity.

Here's how we'll proceed. The two chapters of Part I introduce major problems facing *any* theory attempting to account for life. In Chapter 1 I'll emphasize a philosophical difficulty—the question of *how* we know what we claim to know. The second chapter of Part I throws down the gauntlet. It describes biological systems of astonishing elegance and complexity that demand explanation; many of them were discovered as recently as the new millennium. Part II examines a number of ideas that have been offered as answers, from Darwin's own theory to the most recent non-Darwinian accounts of evolution such as neutral theory and natural genetic engineering. We'll see why, although they may account for some features of life, they all are severely limited in scope.

Part III (Chapters 6 through 9) compiles pertinent evidence from numerous studies on a wide range of species by many insightful investigators. These studies have only become available in the past few decades due to rapid advances in laboratory techniques that closely examine the molecular level of life. The studies indicate that not only is the Darwinian mechanism *de*volutionary; it is also self-limiting—that is, it actively *prevents* evolutionary changes at the biological classification level of family and above. After Part IV (described below), the Appendix reexamines criticisms by top scientists and others of my earlier arguments for intelligent design from the clarifying perspective of more than twenty years later.

The failure of Darwin's mechanism as an explanation for the evolution of all but the lowest levels of biological classification reopens the primordial question of what *does* account for the elegance and complexity of life. My answer appears mainly in Part IV (the final chapter). There I defend the reality of *mind*—a necessary foundation of science itself—and argue that, for its own sake, science must *explicitly* acknowledge mind's existence. Once the reality of mind is affirmed, the explanation for life follows easily. In brief, although chance surely affects superficial aspects of biology, the newest evidence confirms that life is the intended work of a mind and that that work extends much more deeply into life than could previously be seen.

PART I

Problems

CHAPTER 1

The Pretense of Knowledge

The polar bear (*Ursus maritimus*) is the modern world's largest land carnivore, but size and strength don't ensure an easy life. The approximately twenty-five thousand animals are solitary creatures, except of course during mating season when they come together so females can birth an annual litter of one or two cubs. The bears endure dark, bitter winters and perpetually frigid ocean waters as they hunt a diet of chiefly seals. It is a difficult yet majestic role in nature, for which they are superbly adapted.

Ever since its classification as a separate species in 1774, it was realized that the polar bear is closely related to the almost equally huge brown bear (*Ursus arctos*). At first the polar bear was placed in a separate genus. But when it was discovered that the two species could mate successfully, they were both placed, together with the smaller North American black bear (*Ursus americanus*), in the genus *Ursus*. The earliest fossil of a polar bear is over one hundred thousand years old. The species is estimated to have branched off from the brown bear hundreds of thousands of years before that.

Although Charles Darwin didn't mention them in his 1859 masterwork, *On the Origin of Species*, the polar bear is a wonderful illustration of his theory of evolution by random variation and natural selection. Like other examples Darwin did cite, the giant predator is clearly related to a species that occupies an adjacent geographical area, while just as clearly differing from it in a number of inherited traits. It is easy to envision how the polar bear's ancestors might gradually have colonized and adapted to a new environment. Over many generations the lineage could have become lighter in color (making the bears less and less visible to their prey in snowy environments), more resistant to the cold, and more adapted to the sources of food in the Arctic, a process in which each step offered a survival advantage over the previous one.

Yet a pivotal question has lingered over the past century and a half: How *exactly* did that happen? What was going on within the bodies of the ancestors of the modern polar bear that allowed them to survive more effectively in an extreme climate? What was the genetic variation upon which natural selection was acting? Lying hidden deep within the genome of the animal, the answers to those questions were mysteries to both Darwin and subsequent generations of scientists. Only several years ago—only after laboratory techniques were invented that could reliably track changes in species at the level of genes and DNA—was the genetic heritage of the Arctic predator laid bare. The results have turned the idea of evolution topsy-turvy.

The polar bear's most strongly selected mutations—and thus the most important for its survival—occurred in a gene dubbed APOB, which is involved in fat metabolism in mammals, including humans.[1] That itself is not surprising, since the diet of polar bears contains a very large proportion of fat (much higher than in the diet of brown bears) from seal blubber, so we might expect metabolic changes were needed to accommodate it.

But what *precisely* did the changes in polar bear APOB do to it compared to that of other mammals? When the same gene is mutated in humans or mice, studies show it frequently leads to high levels of cholesterol and heart disease. The scientists who studied the polar bear's genome detected multiple mutations in APOB. Since few experiments can be done with grumpy polar bears, they analyzed the changes by computer. They determined that the mutations were very likely to be *damaging*—that is, likely to degrade or destroy the function of the protein that the gene codes for.

A second highly selected gene, LYST, is associated with pigmentation, and changes in it are probably responsible for the blanching of the ancestors' brown fur. Computer analysis of the multiple mutations of the gene showed that they too were almost certainly damaging to its function. In fact, of all the mutations in the seventeen genes that were most highly selected, about half were predicted to damage the function of the respective coded proteins. Furthermore, since most altered genes bore several mutations, only three to six (depending on the method of estimation) out of seventeen genes were free of degrading changes.[2] Put differently, 65 to 83 percent of *helpful, positively selected* genes are estimated to have suffered at least one *damaging* mutation.

It seems, then, that the magnificent *Ursus maritimus* has adjusted to its harsh environment mainly by degrading genes that its ancestors already possessed. Despite its impressive abilities, rather than *e*volving, it has adapted predominantly by *de*volving. What that portends for our conception of evolution is the principal topic of this book.

The Future Starts Now

To understand the profound inadequacy of Darwinism, we must first understand evolution's foundation. Molecules are the basis of physical life. DNA, the carrier of genetic information, is itself a molecule. In turn DNA encodes another class of very complex molecules, proteins, which can join together to form literal machines—molecular trucks, pumps, scanners, and more—that carry out the work of the cell. Among other duties, those machines build the structural materials of everyday life, such as shells, wood, flesh, and bones, which also are all made of particular molecules carefully arranged in particular ways. So in order to more fully understand life, one must understand its molecular basis. The study of the molecular basis of life is the task of my own field, biochemistry.[3]

Because molecules are the basis of life, they are also the basis of evolution. Mutations, the raw material for evolution, are changes in molecules—alterations of DNA and the proteins it codes for. For example, people with the sickle-cell gene have a simple change in their DNA that leads them to produce slightly altered hemoglobin and makes them resistant to malaria. People whose DNA has a small change in a gene dubbed OCA2 lose the ability to produce the molecular pigment melanin in their irises, turning their eyes blue. Most people who hear the word "evolution" probably think of fish with legs or dinosaurs with feathers. Yet they should think of proteins and DNA, because it is molecules that are the underpinnings of visible changes. To more fully understand evolution, one must understand its molecular basis, the biochemical level of life, which we'll explore in subsequent chapters.

Through no fault of his own, Charles Darwin knew none of this. The science of the mid-nineteenth century was primitive compared to today's. The very existence of molecules was still in doubt back then, and the cell, which we now know is filled with sophisticated molecular machinery, was thought to be made of a simple jelly called protoplasm. Perforce the Victorian naturalist was unaware of perhaps the central fact of biology: that heredity—a key prerequisite of his theory—is largely determined by an elaborate molecular code expressed through the intricate actions of hugely complex molecular machines. In the absence of such knowledge, Darwin hypothesized that hereditary traits were transmitted by nondescript theoretical particles he dubbed "gemmules," which supposedly were shed by all parts of the body and somehow collected in the reproductive organs. Gemmules turned out to be wholly imaginary.

Although its components are often unwittingly conflated, Darwin's theory of evolution is actually an amalgam of a handful of separate ideas, several of which do *not* depend as strongly as others on an understanding of biochemistry. For example, the ideas that life has changed over time and that organisms are related by common descent (both of which were controversial in Darwin's time) are supported by evidence from geology, paleontology, and comparative anatomy. Those parts of his theory have withstood the test of time very well.

The situation is completely different for the parts of his theory that we now know *do* depend profoundly on the nature of the molecular level of life—in particular, for the crucial aspects that propose a *mechanism* for evolution. Those portions of Darwin's theory that address the paramount question, "How in the world could such fantastic biological transformations possibly happen?" have for long years gone essentially untested, because research techniques that could probe the molecular level of life in the required detail were unavailable. Partly as a result, Darwin's proposed mechanism of evolution is more widely questioned today than at any time since the role of DNA in life was discovered. To make up for what it is thought to lack, in the past few decades a number of scientists have proposed sundry alternatives to Darwin's mechanism, such as neutral theory and natural genetic engineering. This book will advance a much different theory.

An understanding of the existing molecular basis of life is necessary for an evaluation of any proposed mechanism of evolution, but by itself is woefully insufficient. In addition to that knowledge, the many ways life can *change* at the molecular level also have to be understood—and then the frequencies of helpful ones must be measured and compared for a huge number of organisms over many generations. For all practical purposes that was impossible to do until very recently, when advanced laboratory equipment and new techniques became available to determine the exact DNA sequence of genomes and other critical molecular details. Only in the past few decades could the adequacy of Darwin's proposed mechanism of evolution even begin to be tested.

To put a point on it, up until quite recently speculations on the topic by even the brightest minds were of no more account than were guesses about Earth's place in the universe before the invention of the telescope. So forget what you've heard about how evolution happened. Only now do we have sufficient data to understand the causes of evolution.

Building a solid foundation for understanding that data does require some work. But it brings the substantial reward of a much better appreciation for the place of humanity, and indeed of all life, in the universe. At a minimum, we need a grasp of the outlines of the history of biology, the strengths and weaknesses of Darwin's theory and modern extensions of it, the latest pertinent research results, and crucial philosophical topics. All of that this book will provide in a way that aims to be accessible to the general reading public. The book's goal is to give readers the scientific and other information needed to confidently conclude for themselves that life was purposely designed.

So let's delve right into it. Our first order of business is one of those crucial philosophical topics that's indispensable for evaluating the relevant data: basic *epistemological* difficulties for Darwin's theory. In other words, how do we know what we think we know about evolution? We'll see compelling reasons to conclude that it is not nearly as well supported as it's often portrayed to be.

Evolution and Economics

The study of evolution has a big economics problem. In his 1974 Nobel Prize lecture in economics Friedrich von Hayek decried the "pretense of knowledge."[14] Governments looked to economists for advice on policy questions, and they eagerly gave it. But in reality no one actually knew how to solve the rampant inflation of the time or other pressing problems. Intricate mathematical models were built that included what were thought to be the most important economic factors, but to little avail. Hayek lamented, "As a profession we [economists] have made a mess of things."

The problem wasn't that economists weren't smart. The problem, thought Hayek, was physics envy. Physics envy is the always disappointed yearning by those in a thoroughly complex field to imitate those in a comparatively simple, wildly successful one. As difficult as physics seems to undergraduates, it deals mostly with inanimate matter and can focus on single variables in splendid isolation. Economics, on the other hand, must consider many interacting factors, including people. Economic results are affected not only by supply and demand, but also by competition, taxes, government regulations, technology, and more. They're also influenced by noneconomic human factors such as culture, education, corruption, innovation, jealousy, ambition, disease, population density, greed, charity, and so on. It is effectively impossible to rigorously isolate one of the myriad influences for study away from all others.

So too for the study of evolution. As University of Chicago evolutionary biologist Jerry Coyne once said with a sigh: "In science's pecking order, evolutionary biology lurks somewhere near the bottom, far closer to phrenology than to physics."[5] To be charitable, let's just say closer to economics. Like economics, biology has to deal with, in Hayek's phrase, "structures of essential complexity." Yet the problem is very much worse for the study of evolution, because it concerns processes—many still largely unknown—that occur at the molecular level over thousands or millions of years, involving not only biological factors, but also geological, meteorological, and even celestial ones. Whatever considerations economic science has that confound the accuracy of its prognostications, evolutionary biology has those that affect its pronouncements with exponentially greater force.

Like economics, much of modern evolutionary biology is also cloaked by a thick pretense of knowledge. Of course, biologists can study fossils, genes, and other data to map the history of life reasonably well, just as economic historians can chart how the financial fortunes of industries and nations rose and fell over time. But even at best that just tells us *what* happened. The sticking point is not so much what happened, but *how*. What *caused* events to unfold as they did? That's the question Charles Darwin had hoped to answer.

"Of Course!"

If the subject matter of evolutionary biology is much more difficult than that of economics, then why are its conclusions often presented to the public as indisputable? That of course is a complicated question, but I think a large part of the answer is that evolutionary biology gets little outside feedback when its theories are going awry. Screaming politicians and failing national banks can prod economists into thinking that maybe, just maybe, their computer models missed an important factor or two. But, although its long-term influence on culture may be profound, the idea of evolution has few consequences for practical daily life. And since its main claims are very difficult to test, it can drift along for a long time by dint of intellectual and social inertia, without pushback. Economic proposals often get roasted in popular newspapers and magazines. Evolutionary ones are routinely applauded. Why look for new ideas when everyone is patting your back?

So many evolutionary stories—fish with wrists, hobbits on islands, a predicted disappearance of males—get so much uncritical, gee-whiz-that's-neat media attention that it can be hard for readers to spot serious problems lurking just under the surface. To raise consciousness before we look much deeper, let's next examine three bright-red danger flags that alert us to claims of evolutionary biology that are just make-believe understanding. We'll focus on the first one in this section, and the others in the following two sections.

Red flag number one: Consider the following sentence from an article in *Scientific American* by a writer who was pondering the question of how people differ from other primates:

> Humans have evolved a sense of self that is unparalleled in its complexity.[6]

Contrast it with this sentence:

> Humans have a sense of self that is unparalleled in its complexity.

Now, what information has been lost by deleting the word "evolved"? There have been no studies demonstrating how evolutionary processes could produce a mind with a sense of self. The entire subject of what the mind even *is* has been controversial for thousands of years, with no resolution in sight. In fact, the word "evolved" in the sentence carries no information. It's just a science-y, content-free salute to the notion that everything about living beings—pointedly including the human mind—simply must have come about by the ordinary evolutionary processes that biologists study.

That territorial imperative to plant Darwin's flag everywhere holds even when the topic descends from the sublime to the ridiculous. From an article by a *New York Times* science writer on the pedestrian matter of how some birds clean their rumps:

> Birds like the silky flycatcher, *Phainopepla nitens*, that are mistletoe specialists have evolved a "waggle dance."[7]

Compare to:

> Birds like the silky flycatcher, *Phainopepla nitens*, that are mistletoe specialists have a "waggle dance."

What information has been lost by leaving out "evolved"? No careful studies have been done documenting, for example, how birds lacking that behavior develop it by random genetic changes to specific neural pathways under some identifiable, measurable selective pressure. The word does no real work. It's pretend knowledge. If you're sensitive to it, as I am, you'll find such gratuitous language used routinely in popular science writing and media pretty much every time the topic of evolution comes up.

But, you might ask, isn't that just a peccadillo of popular media? Aren't professional scientists more rigorous than that? No, the same empty phrases often taint publications for working scientists too. For example, in a very technical journal article discussing cellular protein folding, the author remarks offhandedly:

> Another important constraint is the inability of a cell to tolerate significant amounts of unfolded, nonfunctional protein. As a result, every cell has evolved mechanisms that identify and eliminate misfolded and unassembled proteins.[8]

But in fact we have no actual knowledge of how such sophisticated mechanisms could have come about by evolutionary means. We barely know what changes in modern cellular systems would help or hinder their work. Now, reread the quote, this time leaving out the word "evolved." What knowledge has been lost? None at all.

The habit of reflexively affirming current evolutionary theory is inculcated into new generations of students too. For example, in a section on protein structure in a college biochemistry textbook we read:

> Keep in mind that only a small fraction of the myriads of possible [protein] sequences are likely to have unique stable conformations. Evolution has, of course, selected such sequences for use in biological systems.[9]

Note that jaunty "of course." Yet we don't have anywhere near sufficient experimental evidence for the book's conclusion. The authors' confidence isn't based on empirical knowledge—it's feigned knowledge. An unembellished second sentence would read plainly, "Such sequences are used in biological systems."

Gratuitous affirmations of a dominant theory can mesmerize the unwary. They lull people into assuming that objectively difficult problems don't really matter. That they've been solved already. Or will be solved soon. Or are unimportant. Or something. They actively distract readers from noticing an idea's shortcomings. "Of course," students are effectively prompted, "everyone knows what happened here—right? You'd be blind not to see it—right?" But the complacency isn't the fruit of data or experiments. It comes from the powerful social force of everyone in the group nodding back, "Of course!"

When references to it can be dropped from explanations with no loss of information, when proffered evidence for it boils down to a circle of mutually nodding heads, alarm bells should blare that the theory is a free rider.

The "United Front"

Red flag number two: Although almost all popular media routinely portray Darwin's theory as sure scientific knowledge, a number of evolutionary biologists are casting around for something else. A few years ago the world's leading science journal, *Nature*, published a remarkable exchange between two groups of biologists, one defending Darwinian theory and the other arguing that it should be extensively remodeled or replaced. The anti-Darwin side pointed to new results and new phenomena discovered in older disciplines that have been around since the nineteenth century, such as *developmental biology* (the study of how a single cell develops into a fully formed adult organism), as well as to brand-new fields that weren't even imagined until the past few decades, such as *epigenetics* (the study of how factors other than DNA may control heredity).[10] They contended that the new data necessitates a major rethink of evolutionary theory.

The pro-Darwin side pooh-poohed the arguments, claiming that the novel results will fit just fine under the umbrella of Darwin's theory. It's nice to have defenders, but when a significant number of practitioners in a field grumble that a well-known, thoroughly investigated, 150-year-old theory doesn't fit the new data, then something's seriously wrong.

(Interestingly, the upstarts complained that the other side was "haunted by the spectre of intelligent design" and so were driven to show "a united front to those hostile to science" by defending Darwin against all comers. In other words, the anti-Darwin side thought they were being unfairly included in the righteous shunning that pro-Darwinists had instigated to delegitimize ID proponents such as myself. Friendly fire from the war against heresy, they implied, was hitting the wrong target.)

The undercurrent of unrest is shown most clearly by a bevy of scholarly books published since the turn of the millennium by biologists thoughtfully probing evolutionary theory, none of whom think Darwin's mechanism is the main driver of life.[11] Advertisements for the books often announce candidly that Darwin's theory is deficient. For example, one asserts that "the neo-Darwinian synthesis . . . is inadequate to today's evidence,"[12] while another argues for "the inadequacy of natural selection and adaptation as the only or even the main mode of evolution."[13] ("Inadequate" and "Darwinian" seem to be synonyms these days.) An author of another book wrote that "the biggest mystery about evolution eluded [Darwin's] theory. And he couldn't even get close to solving it."[14]

These authors propose novel solutions that they think might rescue Darwin from the data of modern science. We will discuss several of these new evolutionary ideas in Chapters 4 and 5. For now it's enough to notice that the "united front" many biologists present to "those hostile to [their view of] science"—and also present to the general public, which includes you, dear reader—is a public-relations line. It delimits a pretense of knowledge.

The Principle of Comparative Difficulty

Red flag number three: By far the most telling, this red flag comes from a separate area of biology, nutrition, which, like evolution (and economics), has to deal with many interacting variables. The gist of the problem can be seen in recent stories about the effects of cholesterol on human health, where a government panel decided to lift warnings about eating the long-condemned natural product.[15] It turns out, the experts now say, "For healthy adults, eating foods high in cholesterol may not significantly affect the level of cholesterol in the blood or increase the risk of heart disease." As the old *Saturday Night Live* character Emily Litella (played by the late Gilda Radner) would say, "Never mind."

In 2001 *Science* featured an article, "The Soft Science of Dietary Fat," that argued a low-fat diet is not necessarily healthful and that the public had been misled about the relative merits of fat and starch.[16] A few years ago *Scientific American* declared: "It's time to end the war on salt: the zealous drive by politicians to limit our salt intake has little basis in science."[17] About the same time the Department of Agriculture tossed out its iconic food pyramid for a food plate and juggled several of its recommendations.[18] Recently the USDA was reported to be poised to recommend that Americans eat less meat—not because it's better for individuals, but because it may be better for the environment, which indirectly could affect our health too.[19]

In the 1973 movie *Sleeper*, Woody Allen's character wakes up after two hundred years to find that the rules for healthful eating had changed. Deep fat, steak, cream pies, hot fudge—all are actually good for you, the scientists in his new era assure him. It's funny because it seems so plausible.

Here's the ominous significance for the study of evolution. If it's so difficult to determine what is even helpful or harmful for a thoroughly studied contemporary species, *Homo sapiens*—a species that can be monitored under controlled conditions in rigorous detail in great numbers in real time and can even reply to investigators' questions—then how confident can science be about what modifications to their elaborate biology would help or harm a mute theoretical ancestor species in the misty past? In species that can't be monitored or measured? In species that encountered multitudinous biological and environmental influences over millions of years? The stark answer in almost all cases is, we can't have *any* confidence whatsoever.

The problem can be captured by what I'll call the Principle of Comparative Difficulty:

> *If a task that requires less effort is too difficult to accomplish, then a task that requires more effort necessarily is too.*[20]

If we see that a motivated, well-trained long jumper strains to leap even 20 feet in one turn, we'd be silly to expect him to jump 30 in the next. If economists can't correctly predict the behavior of the stock market over the next few months, it's foolish to expect them to accurately forecast the course of a national economy over the next few decades. If nutritionists can't easily determine how one particular dietary factor affects modern humans, then the claim that biologists know which—if any—of countless environmental factors drove evolutionary changes in innumerable organisms in the distant past is ludicrous.

Application of that brightly illuminating principle at points throughout this book, and particularly in Chapter 9, will allow us to plow through much evolutionary clutter and reframe the elemental question of what accounts for the elegance and complexity of life.

The Hard Limits of Knowledge

So far we have seen three red flags that can pop up in our common experience: (1) the frequent gratuitous attribution of elegant, complex, unexplained biology to (presumably Darwinian) "evolution"; (2) the posing of a "united front" by scientists affirming Darwinian claims to the public, even though many biologists express doubts in private and in technical scientific publications; and (3) the fact that biological studies of topics such as contemporary human nutrition, which would be expected to be much easier than evolutionary studies, run into intractable problems. The first two flags show that social pressure is often used to promote Darwinian conclusions well beyond what scientific data warrant.

The third red flag, if we stop to think about it, by itself should make us quite hesitant to put much stock in grand evolutionary claims. Yet, although it is a terrific reason to question whether we know nearly as much about evolution as we had thought we did, it doesn't really tell us *why* we don't know more. Why can't scientists just work harder to get the answers? To answer that question, let's move past simple experience-based reasons to doubt brash evolutionary claims until we hit an epistemic brick wall. Unlike social factors that influence what we think we know, which can be overcome by reading more widely, there are also hard theoretical limits to empirical knowledge that can't be circumvented. Although the popular image of science is dominated by physics and engineering achievements, the reality in many areas is much closer to weather forecasting (Fig. 1.1).

Over fifty years ago a mathematician named Edward Lorenz wanted to take advantage of those newfangled computers to improve

"In an effort to make our economic reporting and projections more accurate, our resident weatherman will be delivering the economic news."

Figure 1.1. Our understanding of evolution is much less certain than our understanding of either economics or weather forecasting.

weather forecasting. The story goes that Lorenz tried to repeat a computer simulation he had run earlier, but he innocently typed in a value for the atmospheric pressure that was ever so slightly different from the first time.[21] Unexpectedly, he got much different results. Confused, he tried again, this time with the correct exact value for the pressure. Now the computer gave the same results as the initial run. But whenever he varied the number a tiny bit, the results diverged in fitful ways.

Lorenz had discovered the chaos principle—the very sensitive dependence of the results of a complex process on tiny differences in starting conditions. That seemingly modest computer test had far-reaching implications. Because measurements of nature always involve some uncertainty, that means chaotic physical systems are *inherently* unpredictable. Since meteorologists can never account for all the factors that might contribute, Lorenz's results spelled doom for the dream of long-term weather forecasting.

And for much else too. Since Lorenz's initial work, more and more systems have been recognized as subject to chaos, from the orbits of planets in the solar system down to even a single billiard ball repeatedly bouncing off the cushions of an ideal frictionless, pocketless pool table. The greater the number of important details of the system one needs to track and the more rapidly they change, the more quickly unpredictability sets in.

A closely related hard limit on our knowledge is this. Chaos theory requires that, even if the operative causes are just ideal simple natural laws, future states of complex systems can't be predicted. But the reverse *isn't* required. It *doesn't* follow that, if future states of complex systems can't be predicted, then only simple natural laws are operating. Rather, any conjunction of causes is undecidable. For complex systems we encounter only unpredictability. Not only don't we know *where* they will end up; we don't know *why* they end there either.

Biological systems always involve many, many significant details that can change relatively quickly, and evolutionary time scales are very long. Thus the sensitive dependence of complex physical systems on starting conditions has rendered real causal explanations in evolution and many other areas impossible, even in theory. The inescapable conclusion is that, although careful studies may show us what *happened*, as far as the physical mechanism is concerned, we can never know in the necessary detail what *caused* the sweep of life's history.

Levels of Explanation

Nonetheless, we need not despair of obtaining at least some profound knowledge of the cause of life, because *not all causal explanations depend on physical mechanisms.* To bring clarity to the basic epistemic question of what we know and how we know it, in this and the next section I'll classify explanations for increasingly complex physical systems into several levels.

The most straightforward kinds of explanations in science are those I call *regular direct explanations*—"regular," because all the objects of study behave in the same manner; "direct," because the properties of an individual object are directly accounted for (Table 1.1). Examples are Newton's laws of motion and Coulomb's law of electrostatic interaction, which are quantitative, describing with mathematical equations exactly how much of a particular quantity there will be, every time.[22] The next level of scientific explanation I call *regular indirect explanations*. They involve simple statistical descriptions. In these cases all the elements of a system behave in the same general manner, but the behavior of an individual element is essentially unknowable. One example is the behavior of gases described by the ideal gas law.

Other statistical studies, however, reflect much more convoluted behavior. An example from medicine is the link between smoking and lung cancer. Statistical studies can tell us to stay away from smoking, but they can't tell us *why* smoking causes cancer. Before rigorous studies were done demonstrating how elements of smoke could damage DNA, our understanding of the physical causal chain was meager. A textbook example of a good statistical study from evolutionary biology is the frequent appearance of the sickle-cell gene in areas of the world where malaria is prevalent. That tells us the two are somehow associated, yet the biochemical mechanism by which the sickle-cell gene is favored in a malarial environment is still speculative.[23]

We can call the level of smoking in relation to cancer and the sickle-cell gene in relation to malaria *manageably irregular explanations*—"irregular," because the various elements of the supposed causal chain have much different properties from each other; "manageably," because we can at least hope that, with massive effort, the physical link between cause and effect can still be tracked down. This is the level at which the *simplest* real-world evolutionary accounts begin. Within this class we see a precipitous decline in our confidence about what is going on in the system.

Table 1.1. Levels of Explanation

Level	Example	Typical Application
Regular direct	Newton's laws	Motion of a body
Regular indirect	Ideal gas law	Container of gas
Manageably irregular	Statistical association	Smoking and cancer; malaria and sickle-cell gene
Hopelessly irregular	None	Detailed long-term weather forecasting, evolution
Spandrels of intelligence	Side effect of mind	Traffic jams, stock market bubbles
Intelligent causes	Intended effect of mind	Complex machinery

The problem for complex systems gets worse *very* rapidly. The statistical associations between smoking and lung cancer or malaria and the sickle-cell gene are strong. Many other correlations—such as for the effect of dietary cholesterol on human health—are weak or imaginary, and any possible causal relationships byzantine at best.[24] It is somewhere in the level of manageably irregular explanations that we begin to trade real knowledge for a pretense of it. When the correlation becomes too weak, when the actual mechanistic causal chain becomes too nebulous, this level quickly blends into the one that frustrated Edward Lorenz—chaos, where no explanation is possible even in principle. We can call this level—the level containing long-range weather forecasting and grand evolutionary narratives—*hopelessly irregular.*

Intelligent Causes and Their Spandrels

The last level of explanation is intelligent causes, and it is the only rational account for many otherwise inexplicable phenomena.

Intelligent causes are utterly different from the physical causes we have so far discussed and, critically, they are *discerned independently of physical mechanisms*. That means, in the right circumstances, the severe limitations that plague mechanistic explanations can be neatly bypassed. We'll consider intelligent causes in depth in the final chapter.

There is a next-to-last level of explanation, which is a kind of transition between unintelligent and intelligent causes. I call that level *spandrels of intelligence*. "Spandrel" is an architectural term meaning "an approximately triangular surface area between two adjacent arches and the horizontal plane above them."[25] In other words it is the space that has to be filled in with building materials to make a square-topped doorway into a round-topped arch.

In 1979 the late evolutionary biologist Stephen Jay Gould wrote a paper with the biologist Richard Lewontin entitled "The Spandrels of San Marco and the Panglossian Paradigm: A Critique of the Adaptationist Programme." Gould was a prominent critic of Darwinian evolutionary theory, which is sometimes nicknamed *adaptationism*. In Gould's view, adaptationism supposedly implied that *every* feature of an organism had been built up by natural selection as a specific adaptation to the environment. He disagreed. He maintained that some biological attributes were simply byproducts of other features and weren't selected for themselves. As an analogy, Gould pointed to the elegantly painted spaces between arches in Venice's San Marco Basilica. You can't have arches without spandrels, and they had to do something with the spandrels, so they decorated them. But it's a mistake to think that the whole building was constructed to feature the spandrels. So too with many biological features, Gould argued.

The technical literature on complexity divides systems into two broad categories: complex *physical* systems, such as hurricanes and sunspots, and complex *adaptive* systems, which can actively alter themselves in important ways in response to external stimuli. A list of complex adaptive systems includes economies, stock markets, traffic systems, political organizations, the internet, brains, insect colonies, ecosystems, the immune system, and biological evolution.[26] One striking thing about the list is that it contains exclusively either groups in which intelligent agents (either people themselves or machinery designed by people) are critical constituents or groups of other living entities. No purely inanimate systems appear.

The systems containing intelligent agents (such as economies, traffic patterns, the internet) display all sorts of unintended, yet intricate and unpredictable behavior. We can view the convoluted patterns as equivalent to Gould's spandrels, expanding the architectural term to mean any patterns or features—including dynamic ones—that are *unintended byproducts* of other, intended, goals.

A simple example of an unintended byproduct of intentional activity is the wood shavings that fall on the ground near the feet of a man whittling, say, a model ship. The shavings might fall into interesting patterns that depend on the wind speed, humidity, density of the wood used, and so on. A diligent researcher might even put together a computer model of where the chips would tend to fall. Nonetheless, the patterns are spandrels.

Since people are a bustling, active lot, some spandrels themselves are dynamic. For example, humans build cars and roads to get from one place to another. Interesting but unintended byproducts of the purposely built structures are that high-volume traffic often slows down wherever three lanes merge to two and that delays occur after an accident, even on the opposite side of a highway, due to rubberneckers. People organize stock markets to facilitate trade; booms, bubbles, crashes, and busts are spandrels of the dynamic system.

In the order of explanation, architectural spandrels are the result of adjacent arches—the arches are not the result of the spandrels. Similarly, economic markets are the result of people organizing to sell goods, not vice versa. In biology, patterns of mutations are the byproducts of the workings of extremely complex molecular machinery over generations, not necessarily the reverse.

If spandrels don't explain arches, what does? The answer, of course, is the final level of explanation—intelligent causes. An architect designed the arches, an architect whose mind had the power to conceive of and work purposively toward a distant goal. The dot-com boom and bust in the late 1990s and early 2000s was a spandrel—the result of the buying and selling of technology stocks on the NASDAQ. But what explains the technology whose ownership was being traded? What explains a computer operating system? Or the construction of a website? Or a cell phone? The minds of engineers, mathematicians, scientists, and other thinking people, whose intellectual work is utterly irreducible to lower-level causes.

If patterns of mutations in genes are spandrels of the workings of the molecular machinery of the cell over time, what explains the molecular machinery? What explains the cellular codes and languages and programs and signals and copiers and motors and trucks and control systems that modern biology has revealed? As I've argued strongly before and will argue again in Chapter 10, like all other complex functional, purposeful arrangements, the stunning sophistication of the cell is best explained by an intelligent cause.

As academic disciplines, economics and evolutionary biology share many of the same problems, because they both build on the same often unpredictable bedrock—intelligent activity. The other causes we have discussed—regular direct, regular indirect, manageably irregular, and hopelessly irregular—all play a role in both disciplines, but in tangential and inadequate ways. The foundations of both disciplines and many of the difficulties both have in accounting for the adaptive systems they study trace back to intelligence.

To a very large extent, the phenomena actually studied by both economics and evolutionary biology are spandrels. Just as economics seldom attempts to account for the widgets whose trading it studies, evolutionary biology rarely tries to reckon with the molecular machinery that powers life. On the few occasions when it does, the rigor of its best efforts falls far short of, say, nutritional studies of cholesterol, and with correspondingly much less certain results. Yet it presses ahead, sometimes doing a reasonable job accounting for some biological spandrels. The diligence of both disciplines is admirable, but it's not a good bet that they will explain their intelligent foundations any better in the future than they have to date.

Where We Were, Where We Are, and Where We Go from Here

This is the third book I've written examining the adequacy of Darwin's mechanism; the first two were *Darwin's Black Box* and *The Edge of Evolution*. Both of those dealt primarily with the riddle of functional complexity in biology—that is, with the need for multiple parts to cooperate with one another to perform some task. That's been a perennial migraine for Darwin's theory ever since the English biologist St. George Mivart called attention to it at higher levels of biology just a dozen years after publication of the *Origin*. *Darwin's Black Box* was the first to argue that the problem was even more debilitating at the molecular level of life. And despite the consternation of Darwin's modern defenders, the difficulty of functional complexity has only grown much worse in the past twenty years.

This book, however, concentrates on completely unexpected, devastating new problems that could only have come to light after major recent advances in technical methods for probing the molecular level of life. With surpassing irony it turns out that, as with the polar bear, Darwinian evolution proceeds mainly by *damaging* or *breaking* genes, which, counterintuitively, sometimes *helps* survival. In other words, the mechanism is powerfully *de*volutionary. It promotes the rapid *loss* of genetic information. Laboratory experiments, field research, and theoretical studies all forcefully indicate that, as a result, random mutation and natural selection make evolution self-limiting. That is, the very same factors that promote diversity at the simplest levels of biology actively prevent it at more complex ones. *Darwin's mechanism works chiefly by squandering genetic information for short-term gain.*

As you might expect, that leads to a big problem. We've seen in this chapter that the claim that biologists know Darwinian processes caused profound, constructive changes in life over the course of billions of years is preposterous—we barely know why existing species do better or worse in their present environments. Yet, thanks to the hard work of paleontology, geology, and other sciences, we know for sure that changes *did* indeed happen. *Something* must have caused them. Can we discern *anything* about the causes? If we can't know in detail what propelled the appearance of eukaryotic cells, the rise of plants and animals, the development of vision, flight, echolocation, or pretty much any of the intricate wonders of life, can we at least identify *some* level of cause that may at least account for *some* momentous effects we see in the history of life, most especially, as Darwin wrote, "that perfection of structure and coadaptation which justly excites our admiration"?[27] Or does any attempt at understanding yield, like Darwinism, just a pretense of knowledge? Those are questions this book will answer. We'll start in the next chapter by examining a handful of the stunning biological wonders for which *any* theory of life must account.

CHAPTER 2

Fathomless Elegance

Not only does the study of evolution have a big economics-style problem, it also has a huge biology problem. Whatever the limitations of their chosen field, economists don't claim to be able to account for the goods whose trading they study. Crude oil, coffee beans, cell phones, automobiles—the question of how any of those ultimately originated is blissfully bracketed. Did oil come from decaying dinosaurs or from the activity of bacteria far below ground? "Who cares!" snort the economists. Are the minds that conceive of engineering wonders such as cars and computers bequeathed by God, epiphenomena of strictly material brains, local manifestations of some universal mind, or something else? "Not our problem," they say with a grin.

Evolutionary biologists, however, explicitly *do* claim to be able to account for the life whose lineages they trace, including bacteria, coffee beans, the humans who design all manner of gadgets, and everything in between. Despite the formidable difficulties that their discipline shares with economics, as discussed in the first chapter, life scientists promise to go far beyond the hazy generalities of their academic colleagues to explain living systems whose relentless detail puts modern electronics to shame.

The credibility of that claim constitutes a major topic of this book. In order to prepare ourselves to evaluate in later chapters whether proposed theories are up to that immense task, in this chapter we'll look closely at details of a few of life's most recently discovered marvels. We'll see that the uncertainties of predicting the market price of pork bellies pale in comparison to accounting for the pork bellies themselves.

A Long Time Ago

Before we begin our tour of some of life's mechanisms, in this and the following section I'll briefly recount a few highlights of the history of biology. This will help to show that mistaken notions—even about the most basic aspects of the world—have always been a part of science, and that huge surprises have often accompanied the development of new instruments to investigate nature.

For millennia primitive humans lived alongside other animals and plants and surely developed a deep understanding of their lives and rhythms. Yet in at least one important sense early people were utterly ignorant of living nature—they didn't know how it *worked*. Imagine a curious person some five thousand years ago gazing at a wildflower or a rabbit and wondering helplessly, "What *is* that?" Folklore provided the only answers, and they were necessarily vague. Without access to such textbooks as we have today, people on their own could only wonder.

The first recorded instance of someone trying to understand life in a modern sense is attributed to Hippocrates, honored as the "father of medicine." Although he and his students could do little to cure maladies, they did at least describe the symptoms of a variety of conditions and so took the first steps to understanding them. His surviving book, *Aphorisms*, records his experience in pithy statements such as "Sleeping too much is as bad as waking too much" and "One man's meat is another man's poison."[1] Perhaps a start, but much work remained to be done.

Another great figure in the early history of biology is one we don't normally associate with it—the philosopher Aristotle, who is actually called the "father of biology." It turns out that Aristotle was interested in many things, definitely including how nature works. From his home near the Aegean Sea in the fourth century BCE he closely observed aquatic life. His writings were a quantum leap beyond *Aphorisms*.[2]

Aristotle's most important lesson for succeeding naturalists was the insight that, to begin to understand nature, one has to go out and closely *observe* it—systematically and in detail. Yet in that recommendation lie both promise and peril. It is true that many of nature's details can indeed be known and understood just by attentive observation. But it is perilous to assume that our limited human senses are *sufficient* for complete understanding; because they can take us only so far, we might miss much of importance and be misled. For example, Aristotle remarked that tiny baby octopuses were "completely without organization." That turned out to be wholly wrong. Modern biology shows their organization to be profound and exquisite. It's just too small for the naked eye to see.

With only their eyes for tools, through no fault of their own, naturalists were stuck at the surface level of biology for thousands of years. Although dissection allowed some progress in understanding large-scale internal anatomy, it too was often misleading. For example, arteries and veins could be seen in dissected animal bodies. Yet the fact that they connected to each other through tiny capillaries in a closed circulatory system escaped even the great Roman surgeon Galen, who thought blood was pumped out by the heart to sink into the tissues, much as water in irrigation canals in his day sank into the ground. His mistaken ideas were taught for thirteen hundred years.

New Tools

Science advances most rapidly when its tools improve. The big breakthrough for biology came in the seventeenth century with the construction of the first working microscope. Although crude by modern standards, the instrument opened up a hidden, completely unsuspected world to view. Anton van Leeuwenhoek was the first to spot tiny living creatures that he dubbed "animalcules"—single-celled amoebae and bacteria.

With more microscopic study even the supposedly familiar world turned alien. What had been thought to be simple, almost featureless insects turned out to have internal organs, weird compound eyes, and many other strange details. Another microscopist of the age, Robert Hooke, coined the term "cell" for compartments he saw in cork tissue, but their significance escaped him. Marcello Malpighi observed red blood cells coursing through capillaries, confirming that blood recirculates in a closed loop, which finally discredited the venerated Galen and freed biology from his ancient authority.

Despite such stunning results and despite its then unimagined potential for the study of diseases, for complex sociological reasons microscopy went into virtual eclipse for over a hundred years.[3] It staged a comeback in the early nineteenth century when Matthias Schleiden and Theodor Schwann advanced the cell theory of life—that all living plants and animals are composed of cells and their secretions and that the fundamental question of life boils down to the question, "What is a cell?"

As the century advanced, so did microscopy, with a variety of improvements that included better lenses, dyes to stain otherwise transparent tissues, and electric lighting to illuminate samples. The twentieth and now the twenty-first centuries have added their own improvements, enormously increasing the power of microscopy with such inventions as the electron microscope; specific molecular tags, like green fluorescent protein, which permit workers to follow a particular one of the thousands of kinds of cellular proteins in the midst of all the others; and enhancements made possible by lasers and computers.

Many other tools besides microscopes have also been brought to bear on the workings of life. X-ray crystallography and nuclear magnetic resonance (NMR, a forerunner of the clinical MRI) allow visualization by computer models of even single molecules. Cloning permits the study of individual genes in isolation and the manufacture of pure, medically important proteins such as insulin and growth hormone. The polymerase chain reaction can select a target gene out of thousands for modification and study. The phenomenal increase of computing power in recent decades makes it possible to record and analyze billions of nucleotides of a species's DNA sequence. In short, we live in a golden age for the study of biology, where many of the shackles that limited naturalists from Aristotle onward have suddenly been broken by a profusion of powerful instruments and techniques.

So what have our powerful new tools shown us about the physical basis of life? The details of studies fill libraries, but the overarching picture is one of fathomless elegance—a seemingly never-ending parade of sophisticated structures, brilliant organizational arrangements, and well-nigh incomprehensibly complex systems. In fact, stunning breakthroughs come so thick and fast in our times that a big danger is to become jaded—to shrug off the next new discovery in the procession of astounding biological features as if it were just the next new mutant character in the twenty-third installment of the *Star Trek* or *X-Men* movie series.

On the topic of such world-weariness the early twentieth-century English writer G. K. Chesterton trenchantly noted: "[Nursery] tales say that apples were golden only to refresh the forgotten moment when we found that they were green. They make rivers run with wine only to make us remember, for one wild moment, that they run with water."[4] In other words, when the fantastical is commonplace, we often have to be jolted into noticing it. As the rush of biological discoveries becomes a cliché, it's easy to miss their significance. Modern biology shows us that we *are* the nursery-tale X-Men—our bodies endowed with fantastic powers we never suspected. We *are* the fairy-tale Borg of *Star Trek*—our cells run by futuristic nanotechnology far superior to theirs. Real life is more marvelous than nursery tales, and real biology more amazing than science fiction.

For the remainder of this chapter I'll survey a handful of marvels recently discovered in life from the level of organs down to the level of molecules. I won't discuss them here as they relate to the specific problems they pose for Darwin's theory. Rather, I'll just describe their structure and organization, mostly letting the exquisite systems speak for themselves. Don't worry about remembering all the particulars—rather, just *feel* the level of detail needed for the systems to work. But do keep the wonders in the back of your mind for when we see in later chapters what random mutation and selection actually are found to do in nature.

Gearing Up

A kid walking through a meadow on a summer afternoon is likely to meet up with any number of fantastic creatures—butterflies, worms, snakes, caterpillars, and more. One of the more attention-getting is the grasshopper, springing to get out of harm's way when approaching sneakers come too close. Ironically, although helpful for escaping other insects, the ability of the bug to jump so far so fast actually attracts juvenile humans, and it's likely to end up in a glass jar for its troubles.

Many a kid, inspired by the hopper, tries to jump as well as it does, only to face the same disappointing limits of biological reality that Wilbur the pig discovered in E. B. White's children's classic

Charlotte's Web. When Wilbur wanted to spin a web, Charlotte the spider urged him to forget it: "You can't spin a web, Wilbur, and I advise you to put the idea out of your mind. . . . You lack a set of spinnerets, and you lack know-how."[5] To perform amazing feats, you need the right stuff. Pigs can't spin webs and kids can't jump like a grasshopper, because they don't have the needed equipment.

So what does it take to be a great insect jumper? Strong muscles and legs, sure. But it turns out that even grasshoppers don't have all the equipment to be the best. That distinction belongs to a different common group of insects called planthoppers. A recent study discovered a phenomenal secret of one of the juvenile stages of the species *Issus coeleoptratus.* It had been reported a half century ago that strange bumps occurred on the hind legs of young planthoppers, but no purpose had ever been assigned to them. Maybe they were just another example of Stephen Jay Gould's biological spandrels—interesting, but functionless.

Wrong. A pair of British entomologists, armed with sophisticated high-speed video equipment, showed that the bumps are actually the *teeth of gears*[6] (Fig. 2.1). For the planthopper to achieve the high-speed takeoff velocity needed to jump hundreds of times its body length, its hind legs must begin to flex in synchrony very quickly, more quickly than it takes for a full nerve impulse to reach the legs. If one leg is triggered before the other, the insect would lose power and tumble erratically. With the gear teeth engaged and the gears spinning at an astonishing fifty thousand teeth per second, as one leg starts to move, the gear rotation starts the other leg moving as well, and the bug gets maximum power and coordination for its efforts.

Although mechanical devices are a dime a dozen at the molecular and cellular levels of life, the planthopper's equipment is the first example of a (relatively) large, in-your-face, interacting gear system in an animal. In an interview with *National Geographic,* one of the authors of the study was betting it wouldn't be the last: "There is

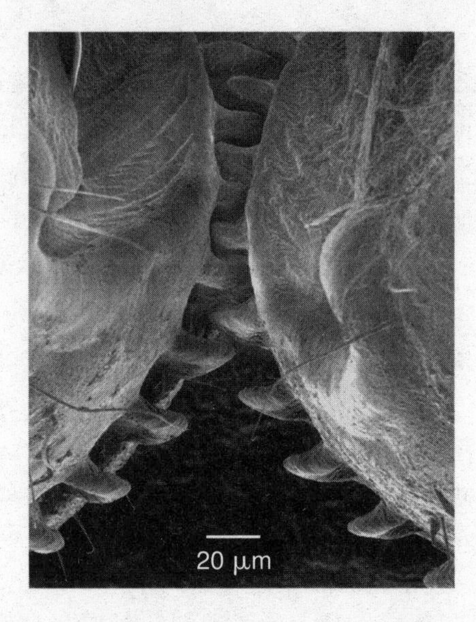

Figure 2.1. Leg gears of the planthopper. The bar marked "20 μm" is less than a thousandth of an inch in length. (From M. Burrows and G. Sutton, "Interacting Gears Synchronize Propulsive Leg Movements in a Jumping Insect," ***Science*** **341 (2013): 1254–56. Reprinted with permission from AAAS.)**

stuff that's vastly more intricate and complicated that hasn't been found yet."[7] Despite the immense recent progress of biology, he is surely right. There seems to be a bottomless supply of wonders in life. Expect plenty more where that came from!

The *National Geographic* writer was thrilled: "This insect has gears. GEARS!" That sound you hear is the jaws of thousands of readers hitting the floor. But why the astonishment? There are plenty of more sophisticated and admirable organs in nature, including the brain—the organ people use to help grasp the significance of all the others. So why is the discovery of gears so sensational? I think one of the reasons is that the purpose is so plain to see. Despite the sophistication of the brain—or maybe because of it—it's not at all clear how the brain works. The same goes even for much simpler systems such as the eye or some molecular machines. It's hard to wrap one's mind completely around their workings, so it's comparatively easy to be talked out of one's strong initial impressions of design, especially by an authority figure in a lab coat. About the gears of *Issus coeleoptratus*, however, there's little ambiguity. The stark clarity of the structure is a standing rebuke to nonpurposive accounts of the system.

The Eyes Have It

The vertebrate eye has been a source of amazement ever since Galen first studied its anatomy in the second century, and it's still going strong. Not surprisingly, in recent centuries it's also been a prominent topic in arguments for and against purposeful design in life. In his 1802 book *Natural Theology*, which began with the famous watchmaker argument for design (which I will discuss in Chapter 3), William Paley went on to admire the workings of the eye, arguing they were even more intricate than a watch, and so pointed even more strongly to a designer.

Like most British college students of his day, Charles Darwin read Paley's book and was quite impressed by it. Later in life he seems to have had it in mind while writing the *Origin of Species*. In a section entitled "Organs of Extreme Perfection and Complication," Darwin agreed with Paley that the eye had many "inimitable contrivances," such as mechanisms to control its focus and correct for chromatic aberration. Yet he also immediately admonished his readers that, although the evolution of the contrivances by his new theory of natural selection acting on random variation may stymie our "imagination," "reason" told him that the difficulty "can hardly be considered real."[8]

It was a clever rhetorical trick: Darwin had actually reversed the roles of reason and imagination. In a different passage he ticked off descriptions of the eyes of different kinds of modern creatures from the relatively simple to the astonishingly complex. But what does that say about the possible evolution of any of them? Darwin's answer was a flight of pure imagination. The passage contains phrases such as "we ought *in imagination* to take a thick layer of [already light sensitive tissue] . . . and *then suppose* [it can vary]"; "we *must suppose* that there is a power [to select variants]"; "we *must suppose* each new state [is reproduced in great numbers]"; and "*may we not believe* [that the process would produce the matchless vertebrate eye]?" (all emphases added).[9]

At best, reason was only a tiny sliver of Darwin's narrative; the vast majority was unfettered imagination. He alluded to a few very broad principles—random variation, natural selection, reproduction, and inheritance—which we can call "reasons." But, as I'll note in the next chapter, all of those principles could be operating furiously and unceasingly and yet lead just to evolution in a small closed loop or to the simplification and degradation of an organism. Innumerable very specific biological details that would have to line up to make his story even feasible resided at a molecular level that was unknown to him. And as we'll see in Chapter 10, contrary to Darwin it is *reason* that tells us that such an intensely purposeful arrangement of parts as we find in the eye indicates design. Unfortunately, for many people in this instance reason seems to be easily overcome by imagination.

In the ever-continuing absence of positive evidence that mutation and selection can make "organs of extreme perfection and complication" like the eye, a kind of negative argument has been offered by Darwin's defenders.[10] The argument is that the eye, although admittedly very impressive, contains a flaw (discussed below) that no designer could ever conceivably have permitted. Since that unequivocally rules out an intelligent agent, perforce the eye arose by an unintelligent process, with Darwin's mechanism being the chief candidate. No need for actual experimental evidence.

Yet there is much wrong with just the logic—let alone the missing science—of the negative argument. For example, even if there were lots of real flaws, "designed" is not a synonym for "unflawed" or "perfect." To see that's true, just ask yourself two questions: Is your car designed? Is it perfect? The two simply don't have much to do with each other. More interestingly, though, recent experimental work shows that the whole negative argument is misbegotten—the supposed flaw is actually a clever feature.

The putative flaw is that, unlike in the otherwise similar eyes of the invertebrate octopus, the vertebrate retina is wired "backward"—its light-sensitive cells are situated in back of the nerve cells that carry an image to the brain. That means light entering the eye has to pass through layers of cells before it hits the retina, which could cause the light to scatter, blurring vision. What's more, in this setup the nerves have to double back through the retina to get to the brain, and the place where they exit the eye has no light-sensitive cells—it's a "blind spot."

The arrangement actually causes no difficulties. The second eye's field of vision covers the so-called blind spot of the first eye, and the brain has clever ways to integrate their visual data. After all, this is the same kind of eye that eagles use to spot small prey from far away—it's magnificently effective. Nonetheless, to design critics it's a "gotcha" argument: no designer would have wired the eye backward, so Darwinism is true; so there.

Even on its own terms the tidy objection began to unravel in 2007 when a team of physicists and biologists showed unexpectedly that light actually doesn't pass through layers of cells to get to the retina.[11] Instead, some cells act as *living fiber-optic cables* to directly channel light from the surface of the structure straight to the rods and cones of the retina (Fig. 2.2). Of course we humans use fiber-optic cables these days in sophisticated telecommunications and computer equipment. Nature does too, but it took until the new millennium for the tools to become available to demonstrate that.

But that's not all. A later study showed that the fiber-optic cables actually improve daytime vision without sacrificing the quality of

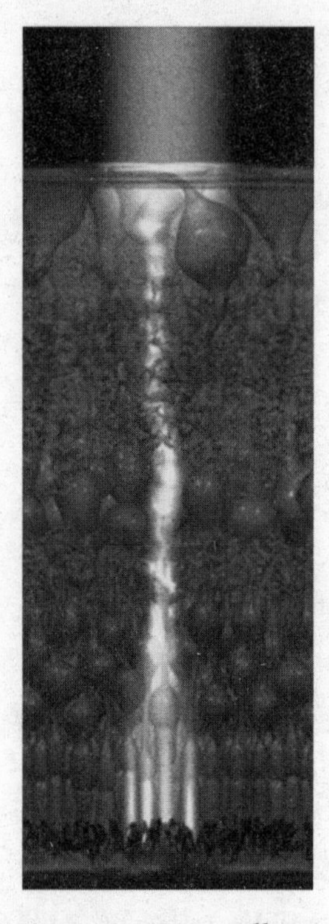

Figure 2.2. Some cells act as living fiber-optic cables to channel light to rod and cone cells in the retina.

nighttime vision.[12] It turns out that each cone cell (used for daylight vision) has its own dedicated fiber-optic cell attached to it, which most efficiently channels wavelengths of light the cones are sensitive to. Light to which rod cells (used mostly for nighttime vision) are sensitive is preferentially released to them by the fiber-optic cables. The science-news website Phys.org could scarcely contain its excitement, exclaiming that situating the photoreceptors behind the retina "is not a design constraint; it is a design feature," and that complaints that it would be better for the vertebrate eye to have its nerve conductors behind the eye, like the octopus does, are "folly."[13]

The fact that octopuses have their eyes arranged differently, in the way that design critics recommend, doesn't make the vertebrate wiring design an error any more than favoring rear-wheel drive in your car makes front-wheel drive an engineering mistake. In fact, the whole reverse wiring criticism is a shining example of the classic logical fallacy called the "argument from ignorance." In a nutshell, the argument goes like this: "We can't think of any good reason for this arrangement; therefore there is no good reason for it. So no intelligent designer would have done it that way." But ignorance of the workings of sophisticated biological machinery is no argument for Darwinism.

The public is often lectured in the most supercilious of tones that Darwin's theory explains all of life, and those who question its ability to account for "organs of extreme perfection" are held up for bullying ridicule. Yet the heavy reliance on no-designer-would-have-done-it-that-way arguments exposes the assertion as rank bluster. Bluntly, Darwinism's icy grip on modern intellectual life is based on shoddy philosophy, not science.

Magnetic Personality

The eye is famously complex and elegant. Let's now turn to something that initially appeared much simpler—until recent research results came pouring in. It's been known for a long time that some migratory birds and other animals could sense the magnetic field of the earth and use it to help navigate around the globe. Yet it's only been about forty years since the same magnetism-sensing ability was discovered in lowly single-celled bacteria. In 1975 the microbiologist Richard Blakemore noticed that bacteria he collected from the bottom of waters near his workplace at Woods Hole Oceanographic Institution in Massachusetts would all swim in the same direction on a microscope slide, unlike, say, the laboratory workhorse *E. coli* (*Escherichia coli*), which darts around every which way. Thinking they were attracted to light, he tried moving the microscope to a darker area, to no effect. But when he placed a strong magnet near the microscope slide, the bacteria changed directions.[14] Blakemore had discovered magnetotactic bacteria, now known to be quite common.

In his pathbreaking paper Blakemore demonstrated that magnetotactic bacteria contain a line of iron-rich particles that later research identified as magnetite—a mineral also found in lodestones (Fig. 2.3, *top*). Having the particles in a line makes their magnetic properties add up, so that they act like one magnet that is strong enough to passively orient the bacteria; even dead bacteria will align with the magnetic field. So what could be simpler to explain? If

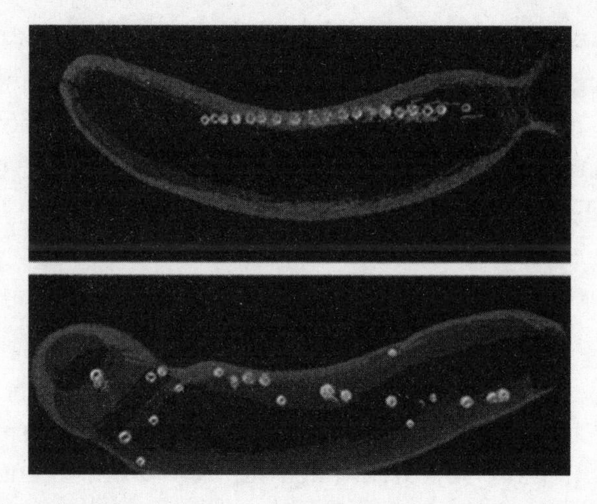

Figure 2.3. ***Top:*** **The magnetosome chain requires supporting cell structures to keep it in a line.** ***Bottom:*** **When a gene for supporting material is deleted, magnetosomes are in disarray.**

together the magnetic particles orient bacteria passively, then all the bugs have to do is swallow a few specks of magnetite. The specks align with the magnetic field, the bacteria align with them, and the problem's solved. Right?

Well, no. Research has shown that it's considerably more involved than one might think, because bacteria don't simply ingest intact minerals that happen to be lying around. Rather, they *manufacture* the right size, shape, and kind of material they need, store it in the correct membranous compartment, called a magnetosome, and attach it to the right place. All of that requires sophisticated control mechanisms in the cell to target the right proteins to the right places at the right times throughout changing external circumstances.[15]

The job the cell faces in constructing a so-called magnetosome can be compared to what a fully automated human factory would face in manufacturing a compass. Just to get a feel for the task, let's quickly run through some of the steps the cell takes. The first problem is that the iron needed is toxic to a cell, so the hazardous material must be handled in a separate compartment to keep the rest of the cell safe. The cell forms the compartment by folding in a piece of its membrane to make a little bag. Now, in a sense membranes are a bit like kitchen plastic wrap—they won't automatically take up the shape that's needed; they need to be formed. So specific cellular protein machinery folds the compartment into the needed form.

Once a separate compartment is made, the bacterium has to import iron into it. Often there is iron dissolved in the watery environment. But even so, another machine—a protein pump—is needed to haul it in from the outside and concentrate it in sufficient quantities to make the magnet. What's more, dissolved iron comes in two chemical flavors we can call "+2" and "+3." The relative amounts of those are very sensitive to the presence of oxygen, yet are critical to making magnetite, which needs exactly one +2 for every two +3 irons. To control the ratio, another protein machine that can electrically convert one type of iron to the other is kept at the scene.

The magnet also has to be the right size and shape and be attached to the cell in the right position. Another protein grabs on to dissolved iron to begin forming the magnetite crystal in the compartment. Other proteins coat the growing crystal to ensure it doesn't get bigger than it should. (If it gets too big, multiple separate magnetic "domains" form, which weakens the crystal's net magnetic field.) Still other proteins actively shape the crystal, just as steps have to be taken in a human factory to make sure an intended bar or horseshoe magnet doesn't turn out to be an amorphous blob. Many species of magnetotactic bacteria form cute little cubes of magnetite, like tiny dice; others form bullet-shaped ones. But no species leaves it to chance. The cell also takes care to line up multiple magnetic compartments head to tail (or rather north to south) in a line, tying them in the right orientation with other proteins and anchoring them to the cell's "skeleton," which is made from still other specific protein machinery. If necessary attachment proteins are experimentally deleted, the compartments are jumbled (Fig. 2.3, *bottom*).

Genetic analysis has shown that different species of magnetotactic bacteria share a common chunk of their DNA, termed a "magnetosome island," which contains all the genes needed to produce their internal compasses. Experimental disruption of any of a dozen different genes results in either the severe weakening or complete elimination of the cell's magnetic response.[16]

Magnetotactic bacteria are more difficult to work with in the lab than many other kinds, and the number of scientists investigating them is comparatively small, so much remains mysterious. Yet from what we know already, even such seemingly simple systems as little magnets in bacteria turn out to require magnificently coherent, purposeful processes to make and use them. Further work will not—cannot—make the system less complex. As with every other area of biology, the more we learn about a system, the more sophisticated and elegant we discover it to be.

Making Tracks

Much of life moves itself, and does so in charmingly diverse ways. People walk, kangaroos hop, fish swim, snakes slither, planthoppers jump, birds fly. Even the tiniest life propels itself, and even there science has found beguiling variety. The best studied of bacterial forms of locomotion is the flagellum, the famous outboard motor that rotates a whiplike propeller at speeds up to 100,000 rpm, allowing tiny cells to zip through liquid as easily as Superman flies through the air. I discussed the bacterial flagellum in earlier books, emphasizing its space-age structure, its mechanical principles, and the severe challenge it poses to Darwinian evolution. But there are other forms of bacterial movement too, and recent research has uncovered a few of their secrets.

A flagellum is fine for swimming, but what if a microbe finds itself on a solid surface with perhaps only a very thin layer of water? In that case bacteria have two general forms of motion, "twitching" and "gliding." As the names imply, twitching bacteria move in discontinuous, jerky motions, while gliding bacteria move smoothly on a surface. Gliding bacteria can be thought of as the unicellular world's answer to snails—both excrete a trail of slime to help them travel along. But snails also use muscles and nerves and other organs that are available only to animals. What do single-cell creatures use to move?

An early conjecture about a gliding bacterium named *Myxococcus xanthus* was that it moved by shooting slime from nozzles at its rear end, relying on the kickback to propel it forward.[17] That would have been enchanting, but it turned out to be wrong. Instead, new information points to something equally fantastic—the cell is essentially a tank that employs a motor to power moving circular treads. Researchers placed a fluorescent tag on a particular protein in the treads of the bacterium, and in the microscope they saw a Day-Glo ribbon that ran the length of the cell, turning as the creature moved along. The authors of the research paper were amazed: "Astoundingly, these helices appeared to rotate within the cell cytoplasm as the cells moved forward."[18] "Astoundingly"—G. K. Chesterton would concur.

Unlike military tanks in our everyday world, the cell isn't heavy enough to generate much friction, so how does the bacterium grip the surface? That's where the slime comes in. Excreted polysaccharide sticks to the surface as well as to the bacterium, giving it traction. Motor proteins use the tread as an internal highway, carrying a load of other proteins whose job apparently is to push against and distort the membrane, making little bumps on the surface that push on the slime, moving the cell forward. The cargo proteins hop off the treads once the motor loses contact with the surface and is circling back. So far at least a dozen proteins are known to be involved in the system. Just think of the controls that have to be in place to make it work successfully.

As if that weren't enough, it turns out there's more than one way for bacteria to glide. A bug called *Flavobacterium johnsoniae*—unrelated to *Myxococcus xanthus*—also glides and also has tank treads. But instead of bumps, it uses proteins that stick out of the cell like tire studs to dig into the slime. In other words, as experts in the field write, the slime "form[s] a 'road' that interacts with specific 'tires.'"[19] The exterior proteins are attached to a baseplate structure under the outer membrane, which in turn is connected to a complex of other proteins. And the complex is indeed very complex. It performs at least two critical functions needed for gliding. First, it is a pump that extrudes the protein pieces that work outside the cell interior—in other words, it helps to build itself.

Second, and most spectacularly, it contains a rotary motor—only the third such motor yet discovered in life.[20] Clever experiments that anchored the cell to a microscope coverslip through its "tire studs" showed the whole cell rotating! It's like fixing the propeller of a plane to a pole and watching as the plane spins around. Since the motor is rotary but the movement of the cell is linear, researchers propose there is a *rack and pinion gear system* to convert the one type of motion to the other.[21] How exactly that might be done remains obscure, but new discoveries will only increase the known complexity of the system.

And on and on it goes. Another, even "simpler," disease-causing bacterium called *Mycoplasma mobile* has a completely different type of locomotion that's been dubbed the "centipede" mechanism.[22] As the name indicates, the bug has a multitude of "legs" that reach out, grip the surface, and pull back. But don't let the biological words fool you—the parts are all hard mechanical features: a lever, hinge, gear, and motor.

Bacterial twitching motion is caused by a pump that shoves rod-like protein material straight out of a tube until it makes contact

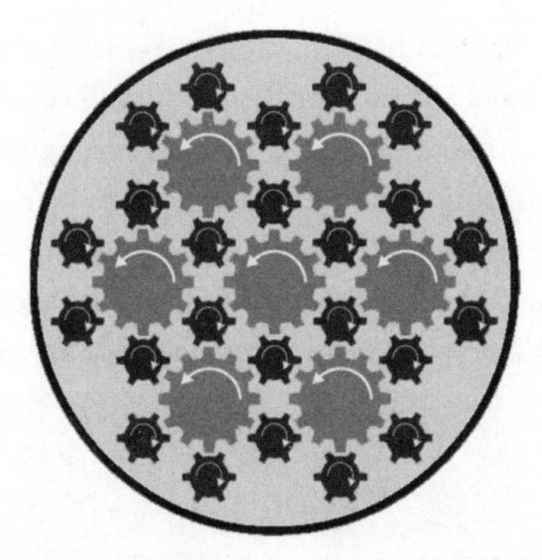

Figure 2.4. Wheels within wheels. Cross section of a proposed model for counterrotating flagellar gears. The larger gears represent closely grouped individual flagella (see Appendix, Fig. A.1, p. 287). The smaller, counterrotating gears represent fibrils that minimize friction. The large circle is the boundary of the structure.

with and grasps the surface. Then the cell mechanically "reels in" the rod, which pulls the bacterium along. A strange group of single-celled microorganisms called archaea have a flagellum that was initially mistaken for the same kind that bacteria have, but recent work shows it's completely different and more powerful.[23] Single-celled eukaryotes—a group different from both bacteria and archaea—swim by an utterly dissimilar means, using "oars" that are "paddled" as in a Roman galley ship; to do so they employ a structure (also, and confusingly, called a "flagellum") composed of hundreds of different kinds of proteins.[24] Bacteria called spirochetes have normal flagella, but they lie inside, not outside, of the cell. These bacteria often have corkscrew shapes that wriggle through water as the flagella spin inside the cell.[25] Using "exquisite architecture" an ocean-dwelling magnetotactic bacterium packs seven flagella together into a narrow tube with matching counterrotating gears to make a turbocharged engine that propels the bug along at ten times the speed of a normal flagellum[26] (Fig. 2.4).

The more we learn of life, the more we realize that *any* type of purposeful motion needs multiple complex well-coordinated parts. At first naive glance under a basic microscope, twitching or gliding or spiraling of tiny bacteria may appear to be simple. Careful investigation, however, reveals a very different state of affairs. Anyone who is amazed at the planthopper's leg gears should faint with shock at the sophisticated engineering of such humble bacteria.

In Control

The elegance of the machinery of life described in the preceding four sections is relatively easy to appreciate once we read about it, because we ourselves use complex machinery every day. More subtle and easier to miss, however, is the need for the very detailed *regulation* of automated machines. When we mow our lawns or drive our cars, it is we intelligent agents who decide where to go, when to turn, and much more. That's easy to overlook. Yet precise regulation must somehow be built into the very structures of biological subsystems. The surprisingly deep regulation of the cell is one of the most active areas of current biological research. This final example section gives a glimpse of it.

Life must actively control its environment. In the absence of control, plants and animals would die, and the surface of the earth would be as barren as Mars. In the early 1960s two French biologists named Jacques Monod and François Jacob took a big first step toward discovering how life exercises control. They examined the ability of the common bacterium *E. coli* to feed on different kinds of sugars. The bug happily ate glucose or the milk sugar lactose if either was present alone in a nutrient broth. But when they were mixed together, *E. coli* would eat exclusively glucose first. Only when that ran out would it switch to lactose. How did a simple bacterium exercise such precise control?

The French scientists proposed a model whereby a length of "regulatory" DNA lay right next to the genes that code for the proteins needed for the metabolism of lactose. When no lactose is present in the broth, a separate gene makes a control protein called a repressor that then binds tightly and specifically to the regulatory DNA, which physically blocks access to the site by a polymerase (a protein machine that makes an RNA copy of the DNA gene). When lactose (or a related metabolite) is present, it binds to the repressor, forcing it to change shape, which causes it to lose its grip on the regulatory DNA so the polymerase can bind. The polymerase, however, can't start working until another protein—call it the activator—binds to a spot on the regulatory DNA next to it. Yet the activator by itself doesn't have the right shape to bind to the DNA. Only when glucose is depleted in the cell will another metabolite (abbreviated cAMP) appear, bind to the activator, and shift it into the right shape to bind to the DNA, which turns on the polymerase, allowing it to begin its work.

Whew! So the genes to make the lactose-metabolizing proteins are only turned on when two conditions are met: lactose must be present, and glucose must be absent. This Rube Goldberg–ish system may sound complicated, but it's actually one of the simplest of genetic control systems. When Monod and Jacob won the Nobel Prize in 1965, hopes were high that their model would explain how all organisms controlled their DNA. Alas, it was not to be. Although the genes of most bacteria, which are prokaryotes, do behave much like those of *E. coli*, those of eukaryotes do not. Prokaryotes are cells without nuclei, and the two main types of prokaryotes are archaea and bacteria. Everything else—from yeast through insects up to mammals—is composed of eukaryotes, cells with membrane-bound nuclei. The gene regulatory systems of eukaryotes surpass those of bacteria like a supercomputer surpasses a slide rule.

Arguably, the first hint that biologists were staring into the maw of unimagined complexity came in 1977. That's when Phillip Sharp and Richard Roberts independently showed by electron microscopy that, when a single-stranded viral DNA gene was mixed with its complementary cellular RNA transcript, the hybrid RNA/DNA double helix had looped-out sections of plain, single-helix DNA. The results indicated that some sections of the DNA gene were not included in the supposed RNA "copy." Roberts's initial paper was fittingly titled "An Amazing Sequence Arrangement at the 5' Ends of Adenovirus 2 Messenger RNA." (Finding the word "amazing" in the title of a sober scientific research article is, well, amazing.) They had discovered split genes. The information to make a particular protein was not, as dogma then had it, a continuous linear sequence of DNA. Instead, genes came in pieces that had to be stitched together to get rid of intervening sequences.

Subsequent work has shown that the great majority of the genes of plants and animals occur in fragments, in stretches dubbed *exons*. The lengths of DNA between them are called *introns*. Split genes can have anywhere from one to dozens of intron interruptions. With multiple exons present, a further complication arises. In what order should the pieces be spliced together when introns are removed from an RNA copy of the gene? It turns out that, although they're usually stitched together in the order they are found in the DNA, sometimes one or more pieces are skipped, or duplicated, or permuted. Such "alternative splicing" uses the same gene to yield multiple proteins (Fig. 2.5). The record holder is a single gene found in the fruit fly that can yield tens of thousands of different proteins—more proteins than there are independent genes in the fly! Alternative splicing increases the protein-coding capacity of a genome far beyond what had been thought.

Another daunting challenge of splicing is the machinery it requires. Many human genetic diseases result from the failure of cells to splice the right ends together. Without the right equipment, a cell would be able to splice RNA about as well as Wilbur the pig can spin a web. So what equipment is needed to stitch the right pieces together with the needed exquisite accuracy?

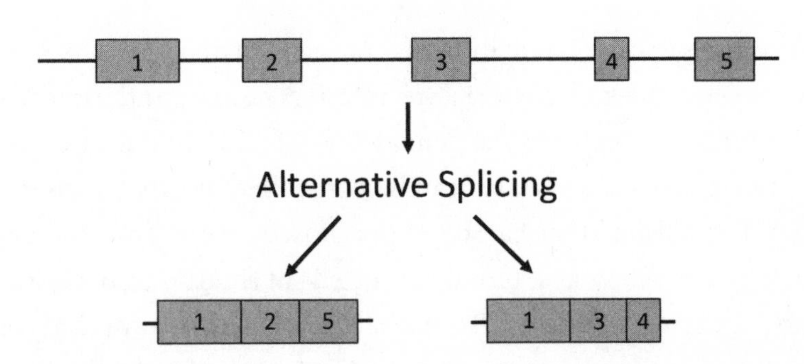

Figure 2.5. Alternative splicing of messenger RNA can yield multiple proteins. The boxes on the top represent exons; the lines connecting them represent introns. Splicing can produce different arrangements of the exons, making different proteins, shown on the bottom.

Although some bacterial introns are capable of splicing themselves, the great majority depend on a supremely complex molecular machine called the spliceosome. The spliceosome consists of a handful of dedicated RNA molecules plus *hundreds* of different kinds of proteins. (For comparison, even sophisticated hemoglobin, which expertly carries oxygen in blood, is comprised of just two different yet somewhat similar kinds of proteins.) The spliceosome is also quite dynamic, with proteins joining and leaving it as needed as it operates.

Since some bare RNAs can splice themselves, clearly not all those proteins are required just to carry out the relatively simple chemical reaction. Rather, they are likely needed to *control* exactly where and when the reaction takes place. Scissors might be enough just to cut a ribbon, but an automated machine that could make fancy paper cutout dolls or artwork would need much more sophisticated cutting tools.

Splicing adds a completely new dimension to gene regulation from what Monod and Jacob proposed. But how about the basic factors they discovered? How about regulatory DNA and protein repressors and activators? Those elements are also found in eukaryotic genomes, but in supercharged forms. The DNA regulatory sequences of bacteria are relatively small and almost always found right next to the genes they control. Eukaryotic genes have those too, but they also can be controlled by DNA sequences called "enhancers" that can be either close (even within the gene itself) or very far away—tens or hundreds of thousands of bases up or down the double helix. To use a literary analogy, prokaryotic sequences are like adjectives that modify a noun they immediately precede. Enhancers are like adjectives that can modify a noun in the next chapter! What's more, although a bacterial gene usually has only one or two protein control factors, eukaryotes can have dozens. It's "the quick brown fox" versus "the quick brown, lithe, crafty, hungry, wiry, . . . [insert many more adjectives] fox." Even in this age of rapid progress, the intricacy of the control elements of eukaryotes still has investigators largely stumped.

How can an enhancer control a gene so far away from it? The trick is that, although the two are separated along the DNA sequence, protein scaffolding has been discovered that folds the DNA, bringing separate regions close together in space. Just as a ribbon can be folded so that two far-separated spots along it may touch, so too with DNA. In fact, it appears that, like a ribbon folded into multiple bows, whole groups of genes are brought close together in discrete spaces inside the nucleus of a cell, so that proteins that need to work together can be made at the same time, and so that the genes can be read more efficiently. To separate different regions from each other, proteins called "insulators" mark the boundaries of the bows. The DNA of a single cell is quite lengthy—several feet for humans—yet is condensed into a microscopic area. The whole operation can be likened to a living pot of cooked spaghetti, where many specific strands have to be brought near each other by tiny machines that cut, fold, and drag the spaghetti pieces to their proper places.

What controls which areas of DNA are brought close to each other? What decides where the boundaries are? What information is needed to direct all the machinery to its proper targets in this dynamic cellular origami? Those and many other questions are all still active areas of research. One safe bet is that regulation will be found to be even more complex than it is now known to be.[27]

Whence Functional Complexity?

Unlike that of, say, a field of boulders at the bottom of a mountain, the complexity of living systems demands explanation because they *do something*, they *work*, they're *functional*. For millennia the eye has been a paradigm of *teleology* in life—something that has an indisputable purpose. Although they have been discovered just recently, the same goes for the other systems described in this chapter—insect gears, tank treads, and the like; the purposeful arrangement of their parts is crystal clear. They aren't exceptions; the more research proceeds, the more and more deeply into life teleology can be seen to penetrate.

So what accounts for the stark purposiveness of the machinery of life? For over 150 years evolutionary biologists have thought they had an explanation—Darwin's theory of evolution. These days, however, as complexity piles upon complexity, a palpable restlessness has gripped the field. In the three chapters of Part II we'll survey the theories that have been offered as accounts, beginning in Chapter 3 with the modernized version of Darwin's own and followed in the subsequent two chapters by newer ideas.

PART II

Theories

CHAPTER 3

Synthesizing Evolution

The *Origin of Species* provoked much excitement at its publication, but by the turn of the twentieth century Darwin's original theory had lost its sparkle. A major reason was that little was known then about the mechanism of heredity, greatly muddying the waters. (I'll have more to say about that later.) However, progress on genetics eventually accelerated and, in the mid-1930s to mid-1940s, led to what is called the *modern evolutionary synthesis* (also known as the *neo-Darwinian synthesis*).

The synthesis brought together separate biological disciplines—genetics, systematics, paleontology, botany, and more—that had rarely talked to each other and so had developed their own peculiar emphases about evolution. Through the holding of meetings and the writing of books, the wandering sheep were guided back into the fold, and a single coherent vision of evolution was forged, which has remained the default view until the present.

In a nutshell, neo-Darwinian theory cites the same basic drivers of evolution that Charles Darwin's original theory did: variation in the members of a species, natural selection acting on that variation, and inheritance of the selected variation by the organism's offspring. The "neo" part comes from incorporating biology that

Darwin hadn't known about: mainly that traits could be inherited through specific, discrete factors called "genes" (which only later were identified with DNA) as well as the mathematics of how those traits would be expected to spread through a population over the generations.

This chapter proceeds as follows. The first three sections discuss Darwin's basic theory and the evidence for it. After a short historical section, the subsequent three sections then illustrate the increased clarity that has come from an understanding of the molecular basis of heredity, which led to the neo-Darwinian synthesis. (Do not get overly distracted by distinctions between the terms "Darwinism" and "neo-Darwinism"; they are often used interchangeably today.) In the last part of the chapter we'll focus on critical yet all too often unnoticed assumptions common to any version of Darwin's theory. All that will give us a solid foundation for understanding the stark challenges to the theories (detailed in later chapters) presented by recent research findings.

The World It Is A-Changin'

One of the most compelling pieces of evidence for an evolutionary view is that the world itself changes. Although that seems pretty obvious to us moderns, it wasn't at all clear to ancient peoples. History tells us, for example, that Aristotle thought the world was eternal and unchanging, remaining over untold eons pretty much as he had found it. In the absence of written historical records and without the easy access to travel and communication we take for granted, people understandably might think that the little patch of ground they call home would never change. Even relatively advanced people are susceptible. Until the big bang theory was first proposed, well into the twentieth century most physicists thought the universe as a whole was changeless.

In the late eighteenth century James Hutton, honored as the "father of geology," also thought that the world showed "no vestige of a beginning, no prospect of an end."[1] But he did think it changed over time and proposed that some geological features could be explained by positing the same forces working over long ages that were known to operate in the modern world. That position, dubbed *uniformitarianism*, was popularized by a later geologist and author, Charles Lyell, who developed uniformitarian theories about volcanoes and earthquakes and the layering of rocks (stratigraphy). Lyell became a good friend of and a strong intellectual influence on Charles Darwin.

Physicists joined the discussion in the late 1800s, when Lord Kelvin proposed that the earth was perhaps a hundred million years old, based on what he calculated would be the cooling rate of the earth's interior. Kelvin didn't know about radioactivity, however, and later scientists showed that that factor bumped up earth's calculated age to 4.5 billion years. If the earth itself has been changing so much over vast ages, then it's compelling to think that life too has changed.

Direct evidence that life itself has changed comes from fossils, which have been known since antiquity. Although he got the age of the earth wrong, Aristotle correctly concluded that fossils are the remains of ancient organisms. Nonetheless, systematic study of the curios had to await the scientific advances of modern civilization.

In the eighteenth and nineteenth centuries great engineering projects exposed layers of previously buried rocks to view. William Smith, an engineer on canal construction projects, noticed that successive layers of rocks contained different kinds of fossils and that the same layer of rock, even if separated from other dig sites by many miles, contained the same kinds of fossils. This led quickly to the thought that different waves of creatures succeeded each other during different periods over the history of life. The French anatomist Georges Cuvier viewed the pattern as *catastrophism*—successive cataclysmic events wiping out much of life repeatedly over the history of the earth, followed by the repopulation of the planet. Charles Darwin, on the other hand, used it as a strong point in his argument for evolution.

The Mockingbird's Tale

Other biological grounds for evolution are the shared traits of organisms, which point to common descent. One compelling—if at first seemingly modest—example of descent with modification came from Darwin's now legendary travels on the HMS *Beagle.* Darwin became intrigued with the birds of the Galápagos Islands, which straddle the equator some five hundred miles west of Ecuador. He noticed that, although South America had only one species of mockingbird, three of the islands of the Galápagos each had its own separate species.

After long ruminating on that and related observations, Darwin conjectured that at some point in the past some South American mockingbirds chanced to land on one of the islands, perhaps blown the considerable distance from the continent by a strong storm. Over time, by the process of variation and natural selection that he would propose, the generations of isolated descendants changed somewhat from each other and from their continental ancestor. If this were true, then species were not immutable, as many people had thought.

Darwin took that idea and ran with it. If the Galápagos mockingbirds had descended from a single South American species, then why not think all the mockingbird species in the world had descended over time from a single ancestral stock by a similar process of isolation in new environments and descent with modification? And if that were the case for mockingbirds, then why not the formation of new species within all other sorts of plants and animals? And if that were the case within all kinds of plants and animals, then why not between more diverse kinds—between, say, mockingbirds and other birds? And if one keeps in mind the great age of the earth and the unceasing variation, competition, and selection that must have occurred, then why not all the organisms on the earth from one ancestral form? "There is grandeur in this view of life,"[2] exulted Darwin.

Darwin had a rival, antithetical explanation in mind as he built his case—separate, individual, immediate creation of each species in its assigned place—and he spared no opportunity to flay it. The flora and fauna of diverse tropical islands resemble those of the closest mainland rather than each other. For example, unlike that of the Galapágos, the biota of the Cape Verde Islands resemble Africa's. Yet if organisms were separately created, asked Darwin, then shouldn't they be matched to the particular geographical conditions they inhabit? Shouldn't all tropical islands have the same life-forms? In the *Origin of Species* Darwin devotes several chapters to the geographical distribution of plants and animals, piling example upon example of similar species living in relative proximity, no matter the environmental conditions, rather than living in distant locations around the world that had similar climates. Separate creation of species for separate particular niches rapidly lost whatever credibility it might once have had.

The Galápagos mockingbirds are just one example from what has grown into the discipline of *biogeography*—the study of the geographic distribution of plants and animals, often with an eye toward discerning evolutionary relationships. Those relationships extend far beyond what Darwin was able to see in his time. Not only are island species similar to those of the nearest mainland, but European and North American species resemble each other closely. Conversely, African and South American species are very different, and the odd organisms of Australia sometimes seem to be in a class by themselves.

These data are difficult to rationalize on just easily observable facts in the present. Instead, a potential resolution had to await the theories of continental drift and plate tectonics in the middle of the twentieth century, a hundred years after Darwin's *Origin*. It's now thought that all the continents of the earth were once joined into a large land mass, Pangea, which broke up beginning two hundred million years ago. Australia separated earliest, South America and Africa later. Europe and North America were in contact as recently as forty million years ago. Perhaps the species of continents that remained in contact longer resemble each other more closely than those that separated earlier, because they've had less time to diverge.

Similarities Between More Distant Groups

The Galápagos mockingbirds seemed compelling, because they were clearly related to each other and to the continental species, yet were noticeably distinct. They were the foot in the door that convincingly broke the principle of the immutability of species. But for his larger theory to be thought true, Darwin had to persuade his readers that there were more similarities between seemingly very different organisms than met the eye.

One of his strongest arguments was from comparative anatomy. The correspondence of the skeletons of mammals had been known well before Darwin wrote. Although much different from each other in proportion, the limbs of widely different creatures—such as horses, bats, and whales—all have the same number of bones in the same relative positions. The anatomist Richard Owen named the phenomenon *homology*, which he defined as "the same organ in different animals under every variety of form and function."[3]

Conspicuously absent from his definition was a *reason* for such an unexpected result. *Why* should structures used for widely different purposes in very diverse environments be so similar? *Why* should the same apparently arbitrary number and arrangement of bones be the best for all of them? Wouldn't a creator have given them all different anatomies, ones that were specially built for their distinctive roles?

Darwin argued that his theory had a ready solution to the puzzle—genealogy.[4] If the diverse creatures had all inherited the basic pentadactyl limb from a distant forebear, and if natural selection had molded it to each creature's needs, then we might expect to see such a pattern today.[5] And if this reasoning were correct, then common descent went well beyond species of mockingbirds, and the power of natural selection must be vast indeed—as vast as the age of the earth.

Not only did the adult skeletons of mammals correspond with each other; so apparently did the developmental patterns of vertebrates, including not only mammals, but birds, fish, and reptiles as well. Embryologists well before Darwin had noticed that the earliest stages of development visible to the naked eye of a wide variety of creatures resembled each other much more closely than did the adult stages.[6] All start out very similar, but diverge in later stages of growth as they assume their adult forms. Darwin argued that both the similarities of embryological forms and the differences of adult forms could be explained as a result of his theory of evolution.[7] Similarities reflected common descent, and differences reflected natural selection for particular environments.

What's more, it isn't only new useful structures that are cited to support the case for Darwin's theory, but old useless ones too—so-called vestigial structures. For example, the appendix of humans is much smaller than it is in other animals, perhaps because it is no longer serving a useful role and thus not conserved by natural selection. Some creatures that live in total darkness, such as cave fish and mole rats, have eyes that can't see due to damaging mutations in genes for functioning eye structures. Why would a creator endow species with useless structures? It's hard to fathom. Yet if the sightless creatures had descended from sighted ones, but over time random mutations destroyed structures that were no longer needed in a changed environment, then the situation would be understandable.

The Need to Revise Darwin's Basic Theory

In this section we'll quickly review the progress of biology in the first half of the twentieth century that required Darwin's basic theory to be modified to explicitly take into account the molecular basis of life. The modified version was dubbed neo-Darwinism.

It's one of history's great ironies that, like everyone else of his time, Charles Darwin had no understanding of the mechanism of heredity (that is, of why offspring resemble their parents)—one of the necessary foundations of his theory of evolution. In the absence of knowledge, Darwin developed a speculative notion he called "pangenesis," in which all parts of the body supposedly contribute tiny particles called "gemmules" that collect in the reproductive organs and somehow carry hereditary information. The theory was completely wrong.

The lack of understanding of heredity caused Darwin's theory of evolution real problems. Based on the observation that offspring often are intermediate in some traits between their parents, a popular notion of the time was "blending inheritance." Biological reproduction worked much the same way as mixing a dark liquid and a light one (or a concentrated fluid and a dilute one); the result is a liquid with intermediate properties. Soon after the publication of the *Origin* an eminent professor of engineering and supporter of the idea of blending inheritance named Fleeming Jenkin reviewed the book and took Darwin to task. How could the small favorable variations that Darwin needed for selection to act upon exert any influence? They would immediately be blended away, diluted over succeeding generations until little remained, wrote Jenkin. In the absence of an understanding of inheritance, the criticism stung.

Meanwhile, even as Jenkin was penning his review, a thousand miles away in the garden of a small monastery in Austria the monk Gregor Mendel was performing studies on hybrid varieties of plants that would launch the science of genetics. Rather than blending, Mendel discovered that several properties of his pea plants like color and texture were inherited intact, undiluted, in neat whole-number ratios. That suggested traits could be passed down more like discrete particles than blending liquids. Unfortunately Mendel's work was published in German in an obscure journal and remained unnoticed until the turn of the twentieth century.

Fast-forward another half century. Increasing economic progress and the availability of electrical power made possible much more sophisticated laboratory equipment. In 1952 Alfred Hershey and Martha Chase showed that when a virus infects and reproduces inside a cell, the virus's DNA enters the cell, but its protein does not. It is DNA (to many scientists' surprise), not protein, that is the genetic material. Around the same time, based on the X-ray diffraction images of Rosalind Franklin, James Watson and Francis Crick showed that DNA was a double helix, and Frederick Sanger determined that proteins had unique amino-acid sequences. A few years later Marshall Nirenberg cracked the genetic code, unlocking the secret of how the sequence of nucleotides in DNA specified the sequence of amino acids in proteins. In another ten years Gobind Khorana and colleagues learned how to synthesize pieces of DNA from laboratory chemicals, Sanger (again![8]) and others invented quick and easy methods to determine the sequence of DNA, and the modern molecular biological revolution was off to the races.

Comparing Sequences

The discovery that DNA is the carrier of genetic information coupled with the ability to sequence it allowed Darwin's theory to be examined at a radically fundamental level that he had known nothing about—the very foundation of life. Like the correlation of continental drift with the similarities of the traits of animals, genetic discoveries undreamt of in the nineteenth century could potentially

Species	Sequence	Differences
human	vlspadktnv kaawgkvgah ageygaeale rmflsfpttk	--
chimp	vlspadktnv kaawgkvgah ageygaeale rmflsfpttk	0
orangutan	vlspadktnv kTawgkvgah agDygaeale rmflsfpttk	2
cow	vlsAadkGnv kaawgkvgGh aAeygaeale rmflsfpttk	4
kangaroo	vlsAadkGHv kaIwgkvgGh ageyAaeGle rTfHsfpttk	9
frog	LlsADdkKHI kaIMPAIAah GDKFgGealY rmfIVNpKtk	22

Figure 3.1. Amino-acid sequence of the first forty positions of the alpha chain of hemoglobin from various species. Each letter is the abbreviation for a different kind of amino acid (*v* for valine, *l* for leucine, etc.). Differences from the human sequence are capitalized. A space is added after each ten letters just to facilitate viewing.

challenge, confirm, or extend Darwin's basic ideas. They've done all three.

Compared to anatomical or embryological data, the ability to sequence DNA and proteins gave science a much more objective measure by which to judge the relatedness of organisms. Instead of just eyeballing the size and shape of various organs and features of one species and subjectively contrasting them to those of another, lengths of exact DNA or protein sequences that corresponded to similar molecular features could be compared—one position at a time, over hundreds or thousands of positions, or even more (Fig. 3.1). As an analogy, it's like trying to judge a person's attitude by either seeing the expression on her face or reading her diary. That dreamy look in her eyes could mean you've caught her fancy, but the diary might paint a completely different picture.

The first results of sequencing of proteins in the late 1950s and early 1960s showed intriguing, unexpected results. Proteins that did the same job were similar yet different between species, but became more different as the biological distance between the species increased. For example, a small protein called cytochrome *c*, which helps produce energy in the cell, was determined to be identical in humans and chimpanzees in all 104 of its amino-acid positions. Between humans and dogs there were 11 differences. Between us and tuna, 21. Between people and moths, almost a third of the total positions differed. Between humans and yeast, almost half.

The differences were quickly interpreted in Darwinian terms as the result of molecular mutations sporadically spreading through populations over long ages—the longer the time since two organisms shared a common ancestor, the greater the number of differences. If that were true, then animals shared an ancestor even with yeast, and all life on earth was likely related.

Completely unexpected was that the number of differences seemed to depend in a regular, clocklike fashion on the number of years the lines leading to the species had been separated (as judged by the fossil record). More puzzling, different proteins apparently "ticked" at different rates, some very fast, some very slow, most in between. The pattern was duly christened the *molecular-clock hypothesis*, and workers struggled to justify why a random, Darwinian process should show such regularity. Ideas were offered and rebutted. After half a century, the idea remains controversial.

Another surprise was that the sequences of different kinds of proteins of the same organism were sometimes similar to each other. For example, the sequences of the two parts of hemoglobin (the protein that carries oxygen in blood), called the alpha chain and the beta chain, are identical in almost half of approximately 140 amino-acid positions. What's more, a simpler protein called myoglobin, which binds oxygen in muscle tissue, is identical with the two chains of hemoglobin in nearly a fifth of its positions. This led to the proposition that perhaps the genes of all three proteins came from an ancient myoglobin-like gene by descent and modification when the original gene accidentally duplicated during the replication of DNA. Many more examples of apparent gene duplication and diversification are known. Thus not only are whole organisms fodder for neo-Darwinian theory; so too are genes.

Comparing sequences of proteins and DNA has allowed progress in judging which species are most closely related. The rough rule of thumb is the fewer differences in sequences, the more closely related; the greater, the less closely related. As with the building of trees of life based on body traits, building trees based on molecular sequences works best for closely related species and gets more difficult as the biological distance increases. For organisms such as bacteria, some biologists have despaired of building a Darwinian tree of life. Comparing sequences does not lead to a consistent tree, perhaps because bacteria have exchanged their DNA over the eons, scrambling any overarching relationships.

The Source of Variation

Up until this point in the chapter we've discussed the great age of the earth and the evidence for common descent as seen in both organisms and molecules. Time plus common descent by themselves, however, don't *even try* to address the most crucial questions about evolution. Common descent is generally invoked to account for *similarities* between creatures, attributing them to a shared ancestor. However, common descent alone explains neither how the ancestor got the traits in the first place nor how the lineages came to differ. In order for Darwin's theory to work, much more is needed.

One essential requirement is that members of a species must vary. So where do new variations come from? That question stumped Darwin, but posterity has provided an answer. Much as the sequence of alphabetic letters carries information in a text such as the *Origin of Species*, the sequence of DNA's four chemical components (called bases or nucleotides, abbreviated A, C, G, and T) carries genetic information. Genetic information is copied by molecular machines in the cell and passed down to an organism's offspring in the processes of reproduction.

As with everything else in this world, those processes, although quite elegant, are not perfect, so occasionally a copying error creeps into the DNA text of the next generation. There are various kinds of errors. One letter may be replaced by another letter, a letter may be left out, an extra letter may be added, or chunks of DNA may be deleted, duplicated, or switched around. DNA from parents is also routinely recombined (that is, pieces are swapped with each other) during sexual reproduction.

The rate at which errors occur can vary, but is about one mistake for every ten billion bases copied or, put another way, for every one to a thousand cell divisions. In bacteria, that means about one every thousand generations. In large animals such as humans, it's about a hundred mutations per generation, because we have much more DNA than bacteria and because there are many cell duplications separating the reproductive cells of one generation from those of the next.

Regions of DNA comprise genes, sort of the way paragraphs are made up of alphabetic letters. The many genes carry the instructions to make thousands of different kinds of proteins, which are the machinery of the cell, performing its necessary tasks such as metabolizing foodstuffs, building structures, regulating processes, and much more. Mutations in DNA can result in altered machinery, affecting the sort of organism that is produced. This is the variation upon which natural selection acts.

Adaptation

Now we arrive at the aspect of evolution that most concerns us in this book. How does Darwin's updated idea explain, as he wrote, "that perfection of structure and coadaptation which justly excites our admiration"? How do the processes he envisioned account for the feathers of birds, the gills of fish, the eye, or any of the marvelously intricate features of life? Or, especially now that science has uncovered the molecular foundations of life, how does neo-Darwinian theory account for the many sophisticated molecular machines that conduct the operations of the cell, such as those described in Chapter 2?

In Darwinian theory an *adaptation* arises because some variations in a population of organisms help its survival. It works like this.[9] Given sufficient food and other resources, any population of organisms would reproduce exponentially until it filled the earth. Yet from observation we see that the numbers of a species usually stay pretty constant over time. Thus there must be a struggle for existence between members of a species for the limited available resources. Since individuals vary in many traits, they will likely differ in their probability of survival. That's natural selection. Then, since many traits are at least partially inherited, the next generation will be enriched in the traits that helped survival. Repeated over many generations, the population changes—it evolves. By variation and selection a species adapts to its environmental niche. In a warmer clime, organisms that have variations that allow them to survive better in heat will be selected; in a drier region, variants that do best there will be selected.

Some theorists, such as Ernst Mayr, one of the founders of the neo-Darwinian synthesis, distinguish between microevolution and macroevolution.[10] *Microevolution* is often regarded as change at or within the level of the species. *Macroevolution* refers to changes at higher levels of biological classification as well as to the appearance of evolutionary novelties, such as each of the forty or so types of animal eyes (all of which appear to be under the control of the same regulatory gene, PAX6). Are micro- and macroevolution due to fundamentally different processes that require different sets of theories to explain them, as some have thought? Or does one meld into the other over time, as others have contended? The point remains contentious.[11]

The history of life is studded with singular, particularly far-reaching events that have affected it in profound ways. It starts with the origin of life itself and continues with the origin of eukaryotes, then multicellular organisms, the development of sense organs, and much more. Yet how do such elegant new biological features arise? Two broad ways that evolutionary novelties have been envisioned to occur, writes Mayr, are by *intensification of function* and *change of function*. In a change of function, a structure that was used for one purpose is adapted to serve a different one; for example, early lungs in fish may have been converted to swim bladders. This is an example of what has been called the "principle of tinkering."

Intensification of function, on the other hand, is exemplified by the eye, which, as Darwin pointed out, has varying structures in different kinds of creatures ranging from a simple light-sensitive spot to the intricate eye of vertebrates. More complex eyes can offer better vision than simple ones. The many kinds of eyes found in nature, it is often asserted, "refute the claim that the gradual evolution of a complex eye is unthinkable."[12] Yet more specific explanations have not been offered.

Making Distinctions

In all-too-brief outline, that is the contemporary case for the theory of evolution as envisioned by Darwin and modified by his intellectual heirs, that all life on earth developed over vast ages by descent with modification, driven primarily by natural selection acting on random variation. It makes for a persuasive story on first hearing and of course has won the support of many scientists in the more than a century and a half since Darwin wrote the *Origin*. Yet over the same time many thoughtful biologists have found it wanting—certainly not completely incorrect, but radically incomplete. And, as we glimpsed in Chapter 1, that restless dissatisfaction is increasing among those who think most deeply about the topic.

How is that possible? How can a venerable, well-studied theory evoke both strong defenders and relentless questions? How can Darwin's idea be, as one scientist commented, "the most important

Table 3.1. The Five Major Concepts of Darwin's Theory of Evolution

1.	The nonconstancy of species (the basic theory of evolution)
2.	The descent of all organisms from constant ancestors (branching evolution)
3.	The gradualness of evolution (no saltations, no discontinuities)
4.	The multiplication of species (the origin of diversity)
5.	Natural selection

intellectual achievement of his time, perhaps of all time" and yet be one that "the biggest mystery about evolution" has eluded?[13]

A large piece of the answer is that "Darwin's theory" isn't just one idea—it's actually a composite of a handful of separate, independent ones. In one of his final books, *What Evolution Is*, Ernst Mayr counted at least five separate concepts in the compound theory (Table 3.1).

Of those five concepts, only two were widely accepted by biologists soon after the *Origin* was published: evolution as such (that is, that life has changed over time) and common descent.[14] Table 3.2 summarizes the views of early evolutionists. Notice that, although other aspects of the theory all found at least some acceptance, Darwin's proposed engine of evolution, natural selection (which then implicitly included random variation—see below), either got none or was thought largely irrelevant.

So it turns out that "the most important intellectual achievement of his time, perhaps of all time" was to persuade virtually all other scientists that life changes along with the maturing earth and that organisms are related by common descent. On the other hand, "the biggest mystery about evolution [that] eluded his theory" was how in the world such a thing could possibly happen. Early biologists were largely unconvinced that major changes in life occurred by selection acting on random variation, and that continues to be a reason for widespread skepticism today.

Table 3.2. Acceptance of Some of Darwin's Theories by Early Evolutionists

	Evolution as Such	Common Descent	Gradualness	Populational Speciation	Natural Selection
Darwin	Yes	Yes	Yes	Yes	Yes
Haeckel	Yes	Yes	Yes	?	In part
Neo-Lamarckians	Yes	Yes	Yes	Yes	No
T. H. Huxley	Yes	Yes	No	No	No
de Vries	Yes	Yes	No	No	No
T. H. Morgan	Yes	Yes	No	No	Unimportant

Darwin's First Theory

The aspects of Darwin's theory (both the original and modern versions) we've just examined are not its only premises. In order to avoid the kind of disabling confusion that all too frequently frustrates discussions of evolution, two other crucial distinctions need to be noticed. To make it easier to follow, let's label the concepts listed in Tables 3.1 and 3.2 as "Darwin's middle theories" and call the two we'll examine over the next few sections "Darwin's first and last theories."

As Darwin often used it, the term *natural selection* really meant natural selection *acting on random variation*. Perhaps because the source of variation was a mystery in Darwin's day and for long thereafter, many scientists then—and even nowadays—seem to unconsciously fold random variation in with selection. Here's an illustration that shows the stark difference between the two. Suppose a biologist grew some bacteria in a Petri dish that contained a mild antibiotic—not enough to kill all the bugs, but enough to substantially slow their growth. Now suppose that same biologist used modern laboratory techniques to add to a second batch of the same bacteria a gene that coded for an antibiotic resistance factor. If she placed those altered bacteria in the Petri dish with the first batch, they would quickly outgrow the originals and take over. Natural selection allows the new bacteria to thrive, but selection is acting on variation that was deliberately added by the lab worker. It didn't arise by random mutation.

Thus there are *two separate* parts to Darwin's mechanism. It is safe to say that virtually no one in science today denies the existence of simple natural selection: if a sufficiently useful variant occurs in a population, probability favors its increase. In the example above of the drug-resistant bacteria, however, the variation on which selection acts is purposefully added to the system. Yet outside of a laboratory, for the provision of the variation upon which selection acts in nature, Darwinism presumes that purpose plays no role. In Darwin's theory, natural selection acts on *random* variation; in neo-Darwinian theory, natural selection acts on *random* mutation.[15] That's the rub. As we'll see in Chapters 4 and 5, almost every biologist who questions the adequacy of the theory doubts the power of *random* mutation.

So Darwin's first theory—the utter randomness of variation—is a much more essential component of his system than any of the others listed in Table 3.1. What did Darwin mean by insisting that variation is due totally to chance? What do later biologists mean by claiming that mutations are random? On a superficial level, later biologists mean simply that changes were not directed toward the good of an animal or species—that the fortunate mutations that led to beneficial variations (and could be gradually built up into such complex systems as the vertebrate eye) occurred by serendipity.

At a profound level, however, Darwin was rejecting teleology—the idea that life is directed toward some end, either by unknown laws of nature, some internal drive, or an intelligent agent external to nature.[16] In the Darwinian lexicon "random" is shorthand for "unguided, unplanned" by anyone—pointedly including God.[17] Virtually all naturalists and philosophers before him thought that nature, and life in particular, was overflowing with purpose. (The topic was called "natural theology.") Darwin explicitly rejected that. As he wrote: "There seems to be no more design in the variability of organic beings and in the action of natural selection, than in the course which the wind blows."[18]

It turns out that an unstated yet fundamental premise of Darwin's entire project—the contention of utter randomness—was based on a bald, simple-minded *theological* assumption: *God wouldn't have done it that way.* A nice benign, indulgent creator wouldn't set up the kind of world Darwin perceived. As he famously worried:

> I cannot persuade myself that a beneficent and omnipotent God would have designedly created the Ichneumonidae [a kind of parasitic wasp] with the express intention of their feeding within the living bodies of caterpillars, or that a cat should play with mice. Not believing this, I see no necessity in the belief that the eye was expressly designed.[19]

Although that prissy view of the deity might have seemed natural to upper-class Victorians tending their gardens in the sunshine, it would surely have surprised the ancient Israelites and religious people throughout history, who routinely endured plagues, persecution, famine, and wild animals. It's strange but true that to a very large degree Charles Darwin insisted the variation that fed natural selection be completely random not because of any actual *scientific evidence* it could suffice, but because of the *theological argument* from evil. Here it is stated as a bare syllogism:

1. If some biological systems cause unnecessary pain, then God did not expressly design any biological system, even the most elegant.
2. Some biological systems cause unnecessary pain.

3. Therefore God did not expressly design any biological system, even the most elegant.

It quickly follows from the syllogism, then, that by default something like Darwin's idea of natural selection acting on random variation simply must be the true explanation. However, if one has any reason to doubt the peculiar premises, then Darwin's whole theological argument collapses, and we're reduced to having to grub for biological evidence to judge what random mutation and natural selection can and cannot do. Delving into that evidence is what we'll do later in this book.

Whistling Past the Graveyard

Another, related *theological* objection that set the stage for Darwin's theory was provoked by the observation that, as more and more species were described in the nineteenth century, more and more amazing contrivances were discovered. Strangely, to some Victorian armchair theologians, that seemed to be a strike *against* God.[20] In their judgment, although no known natural law could explain the specific elegant attributes, there were just *too many* impressive biological features to expect a dignified creator to attend to them all. So neither a world with pain, nor a world with subtle features that can be misinterpreted, such as the eye's blind spot, nor even a world with too much elegance—God isn't permitted to have done it in any of those ways. It seems the gatekeepers hold God on a rather short leash.

That last notion in particular—the notion that there is some identifiable limit on a presumably infinite God's attention to detail; that if there were just a few spectacular biological systems, then, sure, purposeful design by an intelligent agent would be a reasonable explanation, but if there were too many, then somehow God would balk—well, that is a breathtaking deduction. It would seem to require a rather special insight into the mind of the creator, an insight that no one who thus far has made the argument gives any evidence of possessing.

It is certainly correct that there are no general laws of nature that can explain leg gears, flagella, control systems, or the myriad other marvels of life, any more than there are general laws that explain the existence of outboard motors or tanks in our everyday world. Although they're necessary, general laws are woefully insufficient to account for very specific, purposeful arrangements of parts.

The question then becomes, how much confidence should we place in such theological conjectures? Would a designer necessarily be concerned only with the big picture? Or might a designer plan the particulars? Are bacteria too lowly for consideration? Or are they quite elegant? Is attention to detail unworthy? Or is it admirable? What if a designer were indeed quite interested in very many of the physical details of life, the better to assure that any intended goals were reached? If so, then maybe the ongoing discovery of more and more functional arrangements in life—from Aristotle's first close observations of plants and animals, to Robert Hooke's discovery of the compound eyes of insects, to the elucidation of stunningly complex genetic regulatory networks, to whatever further astounding biological features await discovery—is what it so manifestly seems to be: simply the uncovering of more and more of the intended details of life.

Much worse than being dubious theology, however, the dogmatic thinking that from the start rejects a mind behind life also rejects the evidence of our uniform experience—that the purposeful arrangement of parts of a system reliably indicates deliberate design. As I explained in previous books and will discuss in Chapter 10, whenever we see independent pieces ordered to each other to make a coherent whole, we always strongly suspect design. The more pieces there are and the more closely they are matched to the whole, the stronger and stronger is our confidence in the conclusion of intelligent design. Perhaps the most famous illustration of this principle is William Paley's disquisition on finding a watch in a meadow:

> In crossing a heath, suppose I pitched my foot against a *stone*, and were asked how the stone came to be there, I might possibly answer, that for any thing I knew to the contrary it had lain there forever. . . . But suppose I had found a *watch* upon the ground. . . . I should hardly think of the answer which I had before given. . . . For this reason, and for no other, namely, that when we come to inspect the watch, we perceive that its several parts are framed and put together for a purpose. . . . The inference we think is inevitable, that the watch must have had a maker.[21]

Although one has to take care in constructing a valid design argument and Paley admittedly overreaches in places in his watchmaker argument, his main point is exactly correct: *we recognize design in the purposeful arrangement of parts.* We arrive at one of our most basic rational conclusions—that another mind has been at work, that an intelligent cause has been operating—through such observations as Paley described. Although random events surely help shape some aspects of life, we can't draw conclusions about biology from speculative theology. As we'll see throughout this book, the empirical evidence indicates that purposeful design extends very deeply into life.

Darwin's Last Theory

To put into perspective the complete leap into the dark that is Darwin's theory—that a fundamentally random process could produce "perfection of structure and coadaptation"—consider that it was not until more than ninety years after the publication of the *Origin* that observational evidence even showed selection operating in nature. In the 1950s the English naturalist Bernard Kettlewell reported that light varieties of the moth *Biston betularia* resting on tree trunks darkened by industrial pollution in Britain were more frequently eaten by birds. Dark varieties fared much better. The results were hailed as "Darwin's missing evidence."[22] It wasn't until the second half of the twentieth century that the protective power of the sickle-cell gene mutation for malaria was noted, and even later that the damage to the human genome caused by other antimalarial mutations was recognized.[23]

In recent decades many good studies have demonstrated the reality of natural selection. But that was always the easiest part. Who would deny that some features of their biology would affect the survival of organisms? The truly audacious, profoundly nonintuitive, completely unsupported part of Darwinian theory is the almost always tacit, indeed often seemingly unwitting presumption that such a process repeated over time would lead to coherent, integrated, sophisticated, seemingly purposeful systems such as the eye.

And that is the second critical distinction missing from Tables 3.1 and 3.2. Darwin's last theory—call it the "theory of natural coherence"—is the presumption that repeated rounds of random variation and natural selection would, by a succession of separate steps, build elegant compound interactive biological systems. In other words, the claim is not only that undirected evolution occurs or even that it occurs continuously, but that multiple separate rounds somehow come together to form complex organized functional features.

A good example of that unstated assumption comes at the closing of the *Origin*, where Darwin waxes poetic because "from so simple a beginning endless forms most beautiful and most wonderful have been, and are being, evolved."[24] Yet the basic pillars of the theory of evolution—random variation, natural selection, and inheritance—don't say anything about how much an organism might change from its ancestor, let alone that there must be endless forms with beautiful and wonderful features. Without the additional assumption of natural coherence, Darwin's theory is fully compatible with the notion that undirected evolution is restricted to modifying a few preexisting features of an organism in uncoordinated ways.

The situation has not changed at all with time, despite the astounding progress of modern science. Books by even the most distinguished neo-Darwinians, writing after the turn of the millennium and purporting to explain evolution, continue to rely on Darwin's last theory as a bare postulate.[25] No serious evidence is ever presented that selection acting on random variation can, say, convert lungs into swim bladders or produce feathers (or pork bellies), let alone such sophisticated systems as described in Chapter 2. All such assertions rest on the vaguest of concepts, like the previously mentioned "principle of tinkering" or "intensification of function," or on the claim that the evolution of such marvels as the eye isn't "unthinkable" (a standard whose laxity is hard to beat). Rather than serious detailed explanations of how separate tiny changes would accumulate to lead to functional complex systems, which any newcomer to the field would eagerly anticipate hearing, they are treated as unimportant details. After all, the thinking seems to go, we already know the general answer. We know that God wouldn't do it that way, so something like Darwin's theory simply must be true by default. Details to be filled in later, if at all.

Next Up

In Part III we'll examine in detail what Darwin's mechanism has recently been found actually to do in nature independently of such assumptions as Darwin's first and last theories. First, however, in the remaining two chapters of Part II we'll look at extensions to neo-Darwinism that have been proposed in the past few decades by a number of thoughtful biologists (such as those mentioned in Chapter 1 as raising a "red flag"). Understanding their proposals will help us come, by the end of the book, to a confident conclusion about whether *any* fundamentally blind mechanism can account for the elegance of life.

CHAPTER 4

Magic Numbers

The classical neo-Darwinian evolutionary synthesis was devised for a time (first half of the twentieth century) when the molecular foundation of life was almost entirely unknown. It sought to account in the most general of terms for the shapes of beaks, the colors of feathers, the distribution of populations in a region, and other broad, visually observable traits.[1] Early twentieth-century workers applied sophisticated mathematics to evolutionary questions in the field of population genetics, but the nature of the "gene" in "genetics" was then a mystery. (A wit once remarked that, of the four possible combinations, the only one missing from the evolutionary literature was *good* mathematics with *good* biology.) It wasn't until the late 1940s—after the neo-Darwinian synthesis had hardened—that the genetic material was confirmed to be DNA. It took additional decades for the structure of DNA and proteins to be elucidated and even more to begin to flesh out the way genes are regulated.

The jaw-dropping surprises that research has turned up over the past sixty years have left a raft of biologists scratching their heads over how to shoehorn them into neo-Darwinian theory. Many have concluded the surprises don't fit and that a new or extensively revised theory is needed—a "postmodern" evolutionary synthesis, as one quipster remarked.[2] The past several decades have seen a number of proposals to revise or extend neo-Darwinism. In this chapter and the next, we'll survey the most prominent of them, the more mathematically based ones (neutral theory, speculations about a multiverse, complexity theory, and self-organization theories) in this chapter and the more descriptive ones in the next. As we'll see, although the new theories are clever and erudite, and although they do sometimes account for some interesting evolutionary spandrels, none of them even try to grapple with the central problem of evolution that Darwin sought to explain: "that perfection of structure and coadaptation which justly excites our admiration." Modern or postmodern, none account for life's many profoundly purposeful arrangements of parts.

Before we begin, I should mention that this chapter will necessarily be a bit abstruse, because we'll be dealing with abstract mathematical proposals and theory. But I'll use simple analogies as liberally as possible. And if some sections seem a bit fuzzy on first reading, don't worry—there's no exam. You can skim those parts, or skip them entirely and come back later if you want. Rest assured, although the details can be a little confusing, the overarching ideas are pretty straightforward.

Neutral Theory

Charles Darwin didn't know where biological variation came from, and he gave it little thought. Instead, the critical factor in his theory of evolution was natural selection.[3] Selection, he proposed, vigilantly sifts all variations, favoring the good ones and even the slightly helpful ones and rejecting the bad ones and even the somewhat harmful ones. Thus a spectrum of mutational effects is possible from good (such as the otherwise regrettable loss of horns and tusks in trophy animals, which actually increases the animals' chances of survival[4]) to bad. But what happens to those alterations that are smack in the middle? What if a change neither helps nor hurts?

In that case the variation is called neutral and by definition it is invisible to natural selection. Although the concept of neutral variation is implicit in Darwin's description, he gave little space to discussing the topic.[5] He focused on convincing readers that selection could positively adapt organisms to their surroundings. What's more, the sorts of changes that nineteenth-century biologists could study weren't likely to be neutral. They were typically substantive—hefty alterations of anatomy, coloration, behavior, and so on—that would almost certainly affect the organism's chances of surviving.

That began to change when the molecular level of life became accessible to biological investigation. As noted in Chapter 3, methods to sequence proteins weren't developed until the 1950s; efficient DNA sequencing techniques first became available in the 1970s. It's easy to imagine, with Darwin, that a change in, say, the color of a bird's feathers or the length of a bear's fur might help or harm the mutant animal in its struggle to survive. But what about a change in the chain of amino-acid letters of a short region of hemoglobin from `ktnvkaawgk` in chimps to `ktnvktawgk` in orangutans? Does that alteration of the first `a` (alanine) to a `t` (threonine) matter much? There are hundreds of amino-acid positions in the two chains of hemoglobin—do all of them influence survival? It seems unlikely that the job of hemoglobin—carrying oxygen from the lungs to the tissues—is any different in orangutans than in chimps or that the optimum structure of hemoglobin for that role is different in the two species. If both chimps and orangutans do just fine with slightly different hemoglobins, then maybe that change doesn't matter: it's neutral.

The conundrum sharpened when DNA sequencing came to the fore. Proteins are the machinery that takes care of the business of life, so genes that code for proteins are indisputably important. Yet, although all proteins are coded by genes in DNA, the great majority of the DNA of humans and other multicellular creatures does not code for proteins—as much as 99 percent! So what does that non-coding DNA do? Some certainly helps to regulate the protein-coding genes, turning them on and off at the proper time. But it's hard to think that all of it has a definite role. If not, it may not matter if one or even many nucleotide units of noncoding, nonregulatory DNA out of the billions of such nucleotides in mammals are mutated.

The *neutral theory* of evolution began to be developed in the 1960s.[6] It was based on the premise that the very large majority of mutations at the molecular level have no effect on survival. It happily agreed that natural selection determined the course of adaptive evolution, but insisted that only a tiny portion of changes in DNA are in fact adaptive. At the time, the claim that the great majority of mutations that become common in nature have no effect one way or the other on the survival of an organism knocked many Darwinists for a loop. Although Darwin's theory implicitly anticipates neutral changes, few researchers thought that the largest portion of discovered genetic alterations would be immune to what Darwin had touted as the intense, unrelenting gaze of natural selection. The image of natural selection began to slip from that of an eagle-eyed art aficionado, "daily and hourly scrutinising"[7] new work for the most elegant, to a myopic one, who buys many more mediocre pieces than pleasing ones. As we'll see in this and the next few sections, neutral theory is an important advance in our understanding of evolution at the molecular level, a level unknown to Darwin, but even in principle it cannot explain how sophisticated functional features of life arose.

Neutral mutations greatly increase the role of dumb luck in evolution. If a variation is helpful, selection reliably increases it.[8] If harmful, selection surely gets rid of it. But the fate of a neutral mutation depends on many rolls of the dice. For example, suppose that one fly in a population of a million flies was born with a boring neutral change (say, a DNA nucleotide switch from an A to a T at a particular unimportant position), which by definition has no influence whatsoever on its survival. That fly later competes with others of its species to leave offspring. If it's successful, fine—some of its heirs inherit the change (and some don't; by chance they get a copy of the fly's other, unmutated, chromosome). If not, then the mutation is eliminated. Since the fly is initially outnumbered a million to one, the odds turn out to be a million to one that, after millions of generations, the new mutation it carries will eventually (in the jargon of the field) "drift" along by chance to be inherited by all the flies in the species—that is, to be "fixed" in the population. Conversely, there's a 99.9999 percent chance the mutation will be lost before then.

Even though the odds are overwhelming that a particular neutral mutation will be lost, there are many possible positions in DNA that can change. And because mistakes are continually being made by cells copying their DNA over generations, all sorts of neutral mutations are filtering through populations of all species all the time. Neutral theory predicts that the number of neutral mutations that fix in the genome of a species per generation should be constant (regardless of the number of organisms in the species) and is equal to the average number of new mutations that arise in each and every newly born organism. For people, that's roughly ten to a hundred—each and every generation.[9]

Although you might not think so, neutral evolution at the protein and DNA levels can give scientists who study it a lot of information. For example, since the number of neutral mutations that become fixed in a species per generation is pretty constant, then for a species that split into two separate lines of descent in the past, the number of differences between the sibling species will be roughly proportional to the time since they diverged. In other words, the number of mutations is a kind of molecular clock (although a number of factors complicate the analysis).

What's more, comparing the number of mutations between two species in DNA sites that are thought to be neutral with sites that may be functional can sometimes show if natural selection is acting on them. Mutations in neutral sites are expected to accumulate more slowly than in sites where selection is favoring change, and the difference can tell workers if, say, recent changes in a particular protein have helped a species adapt to its environment.

What studying neutral evolution *can't* do is tell us what caused an organism to adapt—the most critical question in this book—since, by definition, neutral mutations have no effect on a species's survival. In fact, since neutral mutations are the bulk of changes at the molecular level, they substantially obscure evolutionary changes that do affect species. For example, the number of differences in the sequences between the hemoglobin of fish and the hemoglobin of mammals is about what is expected from neutral theory. Yet surely a protein that extracts oxygen from water will have at least some differences in its optimum structure from one that extracts oxygen from air. Without further painstaking studies, the few changes that are functional are veiled by the many that are neutral.

Now It's Neutral, Now It's Not

The efficiency of natural selection depends on the number of competing organisms.[10] As a consequence, it turns out that mutations can theoretically switch from being favorable or unfavorable to neutral and back again depending solely on the population numbers of a species. In recent years a few prominent biologists, such as Michael Lynch, of Arizona State University, have argued that the population dependence of such "nearly neutral" mutations might have had profound consequences for the history of life.[11] Without it, they think, the earth might sport nothing but bacteria.

Here's how it works. Bacteria are the most numerous organisms on earth, outnumbering vertebrates by a factor much larger than the number of stars in the Milky Way. Because they have such enormous populations, natural selection is extremely picky about which of their mutations it allows to survive. Now, compared to animals, bacteria have relatively small, very trim genomes. For example, the gut bacterium *E. coli* has only one-thousandth the amount of DNA that a human cell does. And although only about 1 percent of human DNA codes for proteins, almost all of the bacterium's DNA does. What's more, although some of the noncoding DNA of animals does contain sophisticated control sequences, much of it has no known function. In fact, some workers think a lot of noncoding DNA looks like the detritus of ancient viruses that might once have invaded our lineage. Over untold generations, the idea goes, the viral machinery copied itself and added more copies of its genome to the host DNA—that is, to our ancestors' DNA—that was carried along when the host reproduced.

So why is bacterial DNA trim and animal DNA seemingly bloated? Lynch's proposed answer rests on the subtle behavior of nearly neutral DNA. Suppose that a mutation accidentally duplicated a region of functionless bacterial DNA. Perhaps by itself the extra DNA caused no active harm, but since it presents a continuing target that might go on to acquire a mutation that does cause harm, it would be a very small but definite net drawback. Because the population numbers of bacteria are so huge, natural selection would efficiently work against those mutant bacteria with the superfluous DNA. In this scenario the superefficiency actually traps bacteria in a situation where their genomes can't grow any bigger—they are allowed just enough to survive.

Contrast that situation with one in which a relatively small group of cells became isolated from the general population. Now if a mutation increased the size of a bacterium's DNA, inefficient selection would allow it to pass as neutral. If fortune smiles, it can then spread in the isolated group by sheer serendipity. Later, other lucky mutations could occur in the extra DNA to confer some helpful feature—perhaps a regulatory site. Repeat this scenario many times over, and small populations of bacteria could evolve larger and larger genomes with more and more sophisticated features. Eventually, after even more extremely lucky events,[12] they might be transformed into the advanced eukaryotic cells that gave rise to plants and animals.

The point is that it was precisely the *decreased* power of natural selection in the smaller populations that allowed them to acquire more DNA that could then host extra features. So if that idea is correct, then extra DNA arose more through luck than through direct selection.

Will They Come?

The proposed role of neutral theory in the increase of genome size is fascinating, clever, and innovative and may even be correct. Yet it does not even attempt to account for the many functional features that distinguish horrendously complex eukaryotic cells from merely amazingly complex prokaryotes. It may indeed be true that larger amounts of DNA were needed to house the information coding for new molecular machinery and that the supposed obstacles had to be overcome somehow to allow for the extra DNA. But simply adding DNA does not even begin to explain the information that it carries. As an analogy, we can add extra blank pages to a loose-leaf binder to allow for the writing of further chapters of a manuscript. But the blank pages do not explain how an intricate story line comes about.

Typically, after the acrobatics that neutral theory seems useful for explaining have subsided, neutral theorists pass the buck to Darwin to account for adaptive features. Yet, as we saw in the last chapter, neo-Darwinian accounts rely heavily on such vague notions as the "principle of tinkering" and the contention that profound transformations aren't "unthinkable." Leaving aside theological premises, why are so many smart evolutionary biologists so blasé about the evolution of extraordinarily intricate, detailed molecular machinery?

A cogent explanation comes from the eminent mathematical geneticist Masatoshi Nei. Although he is himself a neo-Darwinist, he lambasts grand adaptationist theories as largely speculative even if they are cast in mathematical terms.[13] Nei notes insightfully that, since the molecular basis of mutation was unknown at the time of the evolutionary synthesis, just one-half of Darwin's mechanism—natural selection—was stressed heavily then, and still is today.[14] In other words, even with top-notch biologists, if all you have is a hammer, everything looks like a nail.

Like Darwin himself, most contemporary Darwinists—even ones who are exploring non-Darwinian modes of evolution such as neutral theory—don't worry themselves about the sources of helpful variation. "Constructive mutations" are treated essentially as a foggy, amorphous, undifferentiated theoretical category, and ones needed for the building of complex systems are assumed to be floating around somewhere, available for the asking whenever the selective pressure arises. That works great in a computer model, not so much in the real world.

That same attitude shines through much of the writing of Darwin's staunchest contemporary apologist, Richard Dawkins, who wrote in a dismissive 2007 review of my book *The Edge of Evolution* that the work of the earliest mathematical geneticists indicated "evolutionary rates are not limited by mutation."[15] Yet the great majority of the work of those earlier scientists was done in the first half of the twentieth century, before the molecular foundation of life was understood. Evolution is "not limited by mutation" only if you suppose (as theoreticians often do) that all mutations are equal and that a needed variation for any situation is always lurking somewhere. Like Glendower calling on spirits in Shakespeare's *Henry IV*, all too often neo-Darwinists summon constructive mutations from the vasty deep. But the same question lingers—will they come?

Masatoshi Nei thinks that such mathematical treatments of evolution as Dawkins cites have pretty much been useless in understanding even relatively trivial changes.[16] Although he agrees that natural selection drives the spread of favorable mutations when they're available, Nei argues that the limiting factor that controls the direction of evolution is the availability of *particular* rare mutations—ones that are matched to individual, unique biological circumstances—that may help build complex systems and that natural selection is of distinctly secondary importance.[17] It is mutations, not primarily natural selection, that drive evolution, in the sense that the *right* mutation has to come along at the *right* time to build *particular* cellular systems. As Peter Parker's uncle might say—with great advances in biological understanding of the molecular structures of life comes great responsibility to account for the *details* of how they arose. *Which* favorable mutations were needed? How *exactly* are intricate structures woven together?

Web Spinner

Like the earliest electronic computers, the earliest methods to sequence DNA in the 1970s were slow, clunky, and expensive. And, again like computers, over just decades sequencing technology has become lightning fast, breathtakingly efficient, and very cheap. In 1976 the two-time Nobel Prize winner Fred Sanger and colleagues manually and laboriously determined the complete sequence of a small, viruslike scrap of DNA affectionately named φX174, which is a bit over five *thousand* nucleotides in length. Less than twenty years later, the nearly two-*million*-nucleotide genome of the bacterium *Haemophilus influenzae* was completed. Six years after that, using computer-controlled automated equipment, the three-*billion*-plus-unit genome of humans was solved.

Sequencing was off to the races. Since the turn of the millennium the complete genomes of many familiar creatures have been determined: chimp, dog, cat, cow, bear, elephant, mouse, rat, fish, fly, mosquito, rice, and lots more. Because their genomes are relatively small compared to those of animals (millions rather than billions of units), a much larger number of species of microbes have been sequenced too: bacteria that cause disease and bacteria that promote health; cells that ferment grain and cells that live in hot springs; microbes that can survive on sunlight and microbes that metabolize minerals; viruses endemic to Africa and viruses scooped out of the open ocean.

The enormous amount of raw data generated by the sequencing of all sorts of creatures is stored online, freely accessible to anyone with an internet connection. A new breed of scientists—half biologist, half computer jockey—make a living not primarily by doing lab experiments with microscopes and test tubes, but by using computers to analyze the public data for statistical patterns. Some of those hybrid scientists declare in no uncertain terms that their work shows Darwin was dead wrong—not about evolution in general, but about his image of the tree of life.[18]

Instead of a tree of life, which implies lineal descent of genomes with modification by mutation and natural selection, some biologists advocate a *web* of life. A web model holds that, besides being passed down to their offspring, genes can occasionally also be passed sideways between different species. This concept confuses a lot of people at first until it's explained that this occurs primarily in microorganisms, not larger animals and plants. (After all, even if tiny, microbes are by far the most numerous organisms on the planet—and were the *only* ones for earth's first three billion years.) Biologists think that occasionally some bacteria can engulf other kinds, and some or all of the prey bacterium's genes can be incorporated into the predator's genome. What's more, viruses and other agents appear to be able to shuttle genes between different kinds of microbes. These may be rare events, the argument goes, but they leave their marks in the DNA of the creatures.

By analyzing many microbial genomes by computer, one prominent scientist, Eugene Koonin, of the National Center for Biotechnology Information, has concluded that a lot of genes have been swapped out over the course of evolutionary time, so that it's impossible to trace a single line of descent for an organism.[19] Over long ages all the genes in the microbial world have been mixed and matched sufficiently, so that they seem randomly arranged to him, like the playing cards of a well-shuffled deck. What's more, similar copies of pretty much all genes that exist today, including ones that had been thought to be restricted to eukaryotes, can be traced back to the two primordial groups of prokaryotes—bacteria and archaea.

The fact that all genes seem to have been present from the beginning leads Koonin to downplay evolution after the origin of life.[20] After the origin of life, the idea goes, existing genes were just rearranged in various ways. Koonin strongly agrees with neutral theorists such as Michael Lynch that the eukaryotic cell as a whole—out of which all the familiar plants and animals of our everyday world are composed—and many of its genomic characteristics are primarily the result of neutral drift due to the diminished power of natural selection. After the origin of life, its further evolution is largely just a matter of random currents rearranging the original genes in this way and that.[21] After the origin of life, everything is relatively trivial. After the origin of life, it's all downhill. After the origin of life . . .

So what accounts for that impressive, multitalented origin of life, on which everything else is thought to depend? Hold on to your hat.

Multiverse Theory

Invoking speculative theories of cosmology, bioinformatician Eugene Koonin proposes that we live in an infinite multiverse where any physical event—no matter how unlikely—that is not absolutely forbidden by physical law will happen an infinite number of times.[22] Since an origin of life—complete with all the genes needed for the subsequent unfolding of life as we know it—is not absolutely forbidden, then it has happened by chance repeatedly, endlessly, in some universe or other. Since we find ourselves to be alive here, then we necessarily live in one of those universes where life haphazardly arose.

Koonin is quite serious and sober about his proposal. To show his good faith, he calculates the probability of life arising in a volume the size of our own observable universe and comes to a generous value of 1 in $10^{1,018}$. In other words, he agrees that the odds of life arising even in a universe with life-friendly laws like ours are beyond horrendously bad, well past vanishingly small. Yet, since he takes the multiverse to be infinite, the odds don't matter.

I strongly critiqued the infinite multiverse hypothesis in the final chapter of *The Edge of Evolution*, pointing out its poisonous implications for science or, for that matter, any kind of knowledge about external reality. It leaves us no better off than thinking we're just a brain in a vat. What's more, it's contradicted by the apparent lushness of life, which seems to contain many more sophisticated systems than necessary to produce conscious observers.

Here I'll bypass discussing the idea in depth. Instead, I will just point out that no *explanation* is offered for any functional aspect of life; everything of importance is simply posited as existing from the beginning, the result of one humongous stroke of luck. There is no accounting for properties of systems, no reasoning from patterns, no appeal to processes we see in operation today except to say that, if it weren't this way, we wouldn't be here to make the observation. The entire account truly boils down to the mocking image conjured by physicist and Darwin skeptic Fred Hoyle—of a tornado that passes through a junkyard and assembles a jet plane—except that, to make it "reasonable," Koonin postulates an infinite number of universe-sized tornados and junkyards.

Once you start invoking infinite multiverses to account for elegant biological machinery, it's hard to stop. Koonin uses it not only for the origin of life, but also for the type of irreducibly complex biochemical systems I discussed in *Darwin's Black Box*.[23] The neutral processes (such as we've just discussed) that Koonin calls on to explain the rise of eukaryotes are incompatible with the evolution of complex molecular machines: general eukaryotic features require neutral drift; functional eukaryotic machinery needs strong selection. So, since no actual explanation for them exists, the rationale seems to go, let's all just agree to say they are the result of our good luck in living in the right universe.

Multiverse theory helps not a whit at accounting for life, because it simply posits cosmological unknowns to "explain" biological unknowns. On the other hand, the neutral theory of evolution is a useful elaboration of Darwin's basic idea, which could only have been developed as well as it has after the discovery of the molecular basis of life. It has the potential to mark clear boundaries delimiting where natural selection leaves off and genetic drift begins and to identify genomic features that likely result from the weakness of natural selection. Nonetheless, by definition neutral theory has nothing at all to say about how *sophisticated functional* cellular systems arose.

Complexity Theory

Biology isn't the only intellectual area that's had a revolution in the last half century. Computer science has advanced by leaps and bounds too and in turn has had a big effect on the study of evolution. One way is by allowing researchers to sift ultrahuge amounts of data on DNA and protein sequences for patterns of relatedness. Even though there can be a lot of ambiguities in such studies, the results can often be related to real living systems in a straightforward way.

Another way is much more problematic. Computers have also been used to model evolutionary processes in the hope of discovering hidden features. A computer model of a process is, of course, a mathematical abstraction, not the thing itself, so a perennial danger is that the model doesn't correctly represent the process—that critical but unappreciated details are left out of consideration—yielding misleading results. As a rule, the more complex the system, the very much more difficult it is to build an accurate model. Anyone who lives in a place like Pennsylvania, where the weather can change quickly from day to day, knows that even short-term forecasts ("This Wednesday will bring spotty showers") can be iffy, despite the sophisticated models and advanced computers the National Weather Service has at its disposal.

As we discussed in the first chapter, models of evolution face the same problems as those for weather forecasting or economics, and for the same reason: all depend on a multitude of interacting factors—many not easily measured or even readily apparent—that change quickly with time. Just imagine trying to model the weather in detail over the past hundred million years. Yet that's the scale of the problem that faces grand models of evolution. The mathematical or computer modeling of intricately interacting systems such as weather, economics, and evolution goes by the apt name *complexity theory*.[24]

A favorite approach to complex systems is called *self-organization*. Under the right conditions nonliving matter in nature can organize itself into large, complex phenomena such as whirlpools, tornadoes, and hurricanes. When mixed in the right proportions, some laboratory chemicals form solutions that spontaneously change color periodically or that make regions of different colors in a flat dish.[25] In the living world too fish organize into schools, birds form well-ordered flying patterns, and dispersed cells aggregate in a dish. Perhaps, the thinking goes, the same principles underlie all of these events.

Since interpretable lab experiments on complex systems are notoriously hard to do, from the beginning computer models have often been substituted. An early foray that set the tone for much of the work that followed was a computer program written in 1970 by a mathematician named John Conway. The program was given the evocative title the "Game of Life." To begin the game, a programmer would color squares in a virtual grid (like a crossword puzzle grid) either black or white in whatever pattern he chose. In the next step, the colors of the squares were changed based on the colors of their neighboring squares using arbitrary rules. For example, if a white cell had three black neighbors, it would be changed to black; if a black cell had only one other black neighbor, it would be turned to white. In the next round the colors would be changed again by executing the same rules and the process repeated until it was time for dinner.

When applied to some initial configurations over many steps, some sets of rules would generate interesting, seemingly moving patterns of black squares that reminded some people of little tanks or planes or other things (Fig. 4.1). The idea seems to have been that, since the program produced coherent moving patterns and life produces coherent moving patterns, maybe the program reflected some essential feature of biology.

The Game of Life was fun to play on what were then new personal computers, but exactly what it had to do with life was always obscure. For the reasons we considered in the first chapter, complexity theory is doomed as a real explanation for life's functional systems. Just as the possibility of long-term weather forecasting died with the discovery of chaos, so too did the possibility of seriously modeling the evolution of life.

Self-Organization Theory

For decades some researchers have wondered if self-organization of living systems could somehow extend through time to also account

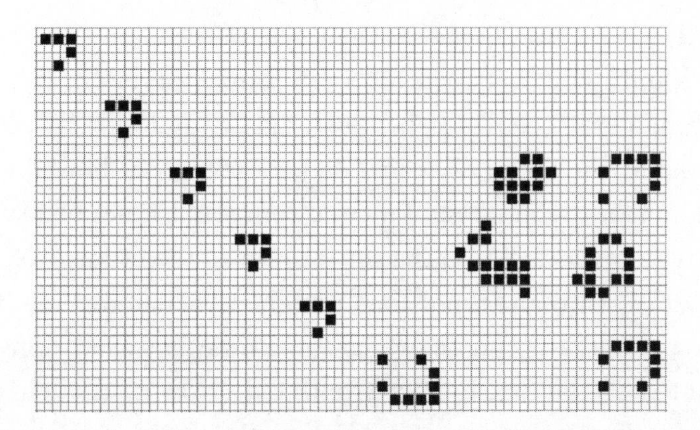

Figure 4.1. A frame from a session of the Game of Life depicting a "space rake" plus five "spaceships." The relevance to biology is not apparent.

for their evolution. In the later twentieth century some mathematically adept biologists began to ask more seriously whether, rather than Darwin's random variation, self-organization might account for the complex systems of life. One of the most prominent was Stuart Kauffman, then of the Santa Fe Institute, a think tank dedicated to exploring complexity theory. Kauffman dubbed his idea "order for free." The gist was that many sufficiently complex systems naturally fall into patterns similar to those seen for real genetic networks in living cells. If that were the case, then maybe genetic networks somehow organized themselves, and no messy random variation or gradual modifications might be needed to account for DNA regulation.

Kauffman envisioned an abstract system in which each of a number, *N*, of components could control one or more other components, switching them between two states such as "on" and "off." The number of other components each controlled was designated *K*, so the systems were called *NK* systems. He showed that if the number of controlled components, *K*, was only one, the network was frozen in a single state that couldn't change over time. If *K* was three or more, the whole network changed capriciously. When the number of inputs, *K*, was two, however, then, like in the Game of Life, the system could switch between a limited number of states.

Kauffman gave the region between rigidity and capriciousness a catchy name, the "edge of chaos." What's more, he showed mathematically that the number of unique connected states of a system in that region was equal to the square root of the number of components, *N*. Kauffman argued that *N* might represent the number of genes in the human genome, which at the time he was working was thought to be about one hundred thousand. (Now it's thought to be a fifth of that.) The square root of 100,000 is a little more than 300, which is approximately the number of different cell types (such as skin, nerve, muscle, blood, etc.) that people have. So, he contended, maybe the genetic program of each cell type corresponded to a unique state of an *NK* system.

In the mid-1990s Kauffman wrote a technical book on *NK* systems, *The Origins of Order*, followed by a popular work on self-organization, *At Home in the Universe*, in which he argued that we should feel like a natural part of the world—we should feel at home—because the behavior of complex systems naturally gives rise to life and beings like ourselves. Unfortunately, the link between the evolution of real genetic networks and *NK* systems is no more apparent than the link between life and the Game of Life.

For biology the study of self-organization is at best a classic case of focusing on spandrels rather than on the thing itself. Just as the study of traffic jams yields little information about how cars are made, the study of flocking birds and schooling fish tells us nothing about how the eyes and nerves and senses and chemical transmitters and receptors and so on that are necessary for those animals to congregate may have arisen. Similarly, the study of *NK* networks says nothing about the structure of the proteins and genes and cells they require. Although studies of self-organization may shed some light on how life *behaves*, they say little to nothing about how life *arose* or *developed*.

Pre-Self-Organization Theory

A more recent book on the modeling of evolution from another member of the Santa Fe Institute is *Arrival of the Fittest: Solving Evolution's Greatest Puzzle* by Andreas Wagner. The "greatest puzzle" of the subtitle is the problem to which Darwin gave little thought but that preoccupies most of his modern critics: How could random mutation produce the right pieces to build such elegant systems as we find in life? Wagner doesn't invoke self-organization exactly in the same way as Stuart Kauffman. Instead of molecules and genes and proteins organizing themselves, Wagner argues that the very space of possibilities available for the rearrangement of all those things is somehow structured to facilitate their productive evolution—in other words, structured to easily move between beneficial states with just a few evolutionary steps.

After introductory sections, Wagner begins his argument by considering metabolic reactions in a large number of different kinds of bacteria. Metabolic reactions are chemical transformations that either break down foodstuffs to simpler molecules (often capturing their energy in the process), build up simpler molecules into the more complex structures of a creature's body, or rearrange molecules into more useful forms. Although there are major common themes, the specific ways in which various molecules are made or degraded can differ in diverse organisms. Like much scientific information these days, the known metabolic reactions for many kinds of bacteria are stored in internet databases, accessible to any researcher who wants to analyze them.

Wagner considers the approximately five thousand individual chemical reactions for all bacteria in the database that are used to make sixty critical molecular components that all cells need (things like amino acids, nucleotides, vitamins, etc.). Some bacterial species use some of the reactions and other species use other ones to arrive at the same critical components. He then asks whether he can theoretically switch genes in his computer for reactions that different bacteria use, one by one, and have a cell remain viable—that is, still able to make all sixty required components. He finds that to a very large extent he can swap out reactions one by one and the cell would theoretically survive. What's more, by changing one reaction at a time, he can begin with any particular bacterial species's metabolic profile and end up with any other species's profile.

Andreas Wagner thinks the results show that the genetic space of metabolic possibilities is somehow mysteriously organized to allow for the diversity seen in the earth's bacteria.[26] I think the results are interesting, but not surprising. Picture a big city like New York, with the street system laid out as a grid. Many buses travel up and down the avenues and back and forth across streets. Suppose you wanted to visit sixty city landmarks. If one of the most direct bus routes wasn't operating the day you wanted to see a certain landmark, could you still get there by public transportation? Almost certainly. You very likely could hop another bus and get there by a different route after a couple of transfers. As long as there were enough routes that crossed each other and passed by the landmarks, an enterprising tourist would have no problem. Since cells catalyze many interrelated reactions and since all of them converge on the same sixty critical components, it's not surprising that there are multiple ways to arrive at each component.

A bigger problem with Wagner's argument, however, is that it doesn't solve evolution's greatest puzzle—it doesn't even try to account for the cellular machinery that is catalyzing the chemical reactions to make the needed components. Like Stuart Kauffman's self-organization, it concentrates on spandrels, not on arches. A modern refinery might have many distillation towers and heating chambers and computer-controlled valves with which it synthesizes various chemicals. A rival company might buy their refinery equipment from a different manufacturer and synthesize the same chemicals using different reactions. But showing that the equipment of the two refineries can be switched out and still make the same products says nothing about where the equipment came from.

Feet Back on the Ground

This chapter has touched on a handful of mathematically based ideas that have been offered in the past few decades as supplements to neo-Darwinism: neutral theory, the effects of a multiverse, complexity theory, and self-organization theories. None of them work. Neutral theory by definition doesn't even try to explain the beneficial complex molecular machinery at the foundation of life. A multiverse is also no explanation, because it just invokes astronomical speculations to wave away biological mysteries, effectively attributing life to one humongous accident. And the connection of neither complexity theory nor self-organization theories to the evolution of the actual machinery of life has ever been elucidated.

Still, more than a few readers may feel a little shaky dealing with math. So this seems like a good time to revisit the Principle of Comparative Difficulty from Chapter 1, wherein we noticed that if an easier task is too difficult to accomplish, then a harder one most certainly is too. Recall that the distinguished evolutionary biologist Masatoshi Nei scorns the utility of mathematical population-genetics theory even for relatively minor short-term real-world evolutionary changes, calling the mathematical approach "practically powerless."[27] Yet if modeling even minor evolutionary effects is quite problematic, then the types of studies done by Stuart Kauffman, Andreas Wagner, and many others—which hope to account for massive evolutionary changes that occur over lengthy time frames—are simply pushing mathematical tools far past what they already labor unsuccessfully to explain. Mathematical models can't explain greater evolutionary changes if they can't account for lesser ones. They yield only a pretense of knowledge.

Like the classical neo-Darwinian disinterest in the details of eye evolution, the newest computer models on offer have little or nothing of substance to say about how the elegant structures of life arose—no planthopper gears, no bacterial tank treads, nothing real. At best, mathematical approaches such as neutral theory try to account for tangential facets of biology. In the presence of magnificent Gothic arches, they all press their noses firmly to the spandrels.

Breathing the thin air of abstract theory for too long can induce hallucinations, and we start to imagine phantasms transforming themselves into whatever we wish to see. To begin shaking ourselves awake, in the next chapter we'll look at some newer ideas about evolution that are at least grounded in real biological features, but nonetheless are hard to fit into classic neo-Darwinian theory. After that we'll fully reconnect with reality. In Part III I'll discuss the results of laboratory experiments and field observations that show what evolution actually does in the real world, independent of our hopes.

CHAPTER 5

Overextended

When I was in college we were taught a programming language called Fortran, which was then preeminent in engineering and science computing. I remember someone asking a computer science professor at the time what he thought the dominant engineering programming language would be like in twenty years. He had no idea, he said, but whatever it was *like*, it would be *called* Fortran.

Once there's an established brand in an area, it's often easier for innovators to keep the label but change the content rather than to persuade people to accept something unabashedly novel. That rationale might be lurking behind the witty comment about the modern evolutionary synthesis being superseded by a "postmodern" evolutionary synthesis. It almost surely motivates the branding of the so-called *extended evolutionary synthesis* (EES).[1] In their writings, proponents of the EES routinely offer the most fulsome praise of Charles Darwin and the most sincere assurances that they don't want to change the modern synthesis at all, not one little bit. They just want to extend it, broaden it, improve it—so that it includes even more factors, ones of which Darwin was totally unaware.

In practice, the EES is a hodgepodge of disparate, partially overlapping observations, concepts, and hunches—some not too far removed from the venerable modern evolutionary synthesis, others pretty far out there—grouped together mostly by a shared dissatisfaction with the status quo. In this chapter I'll describe the most prominent ones called, in order: evo-devo, facilitated variation, inclusive inheritance, niche construction, developmental plasticity, natural genetic engineering, and game theory. (The first two, evo-devo and facilitated variation, I'll just mention briefly, since I wrote about them previously in *The Edge of Evolution*.) As with the math-based ideas discussed in the last chapter, we'll see that the proposals gathered under the umbrella of the EES *don't even try* to explain the complex functional structures of life. This will give us a solid grounding for evaluating the most recent research results, detailed in Part III.

Evo-Devo and Facilitated Variation

Evo-devo is a nickname for *evolutionary developmental biology*. Plain old developmental biology (which arose from embryology) is the study of how creatures grow from a single cell to their adult forms. The discipline used to be entirely descriptive—which cells or organ systems form first, which come next, and so on—until the late nineteenth century. At that point embryologists turned to "developmental mechanics," which sought causal explanations for developmental phenomena. Then, just a few decades ago, advances in understanding the molecular basis of life identified "master" genes and proteins that helped control facets of embryo development. Much to the surprise of workers in the field, many of those genes and proteins turned out to be quite similar between widely different types of creatures. Comparison of the sequences and arrangements of the genes between different species then allowed for conjectures about who descended from whom; hence "evolutionary" developmental biology.

Master genes control cascades of events that can lead to the development of an entire organ or anatomical feature. For example, a widely discussed gene dubbed PAX6 controls the development of varied eyes in animals, including both the camera eye of vertebrates and the compound eye of some invertebrates. Startlingly, when a mouse PAX6 gene was transplanted into a fruit fly, the flies developed ectopic eyes (fruit-fly eyes, not mouse eyes)—eyes in the wrong places, including on their legs and antennae. The organs weren't connected correctly to nerves, so they didn't allow the flies to see with their limbs. Still, it showed that big anatomical consequences could follow from relatively small genetic changes.

Other master genes control other significant aspects of biological development. Yet they aren't magical substances—they are simply switches that activate the downstream machinery they're already connected to. As an analogy, the power button on your computer could be called a "master switch"—it activates everything else. But it gets whatever abilities it has from the way it is wired into the other pieces. A computer technician might remove the power switch from a PC and connect it to a Mac or even to a radio, but the switch can activate the other devices only because they were already wired for action. By itself the switch does not produce anything.

Instead of using solid circuits as computers do, proteins coded by master genes work by traveling through the cell and binding to specific "signature" sequences of DNA that are close to the genes they activate. So if a gene that is *not* already activated by the master gene somehow acquires a copy of the signature sequence, it might then also be turned on when the master gene is triggered. Laboratory workers these days can easily add such a sequence near a gene to study the effects of its activation by the master gene.

EES proponents who focus on evo-devo divide roughly into two groups. The first speculates that, once master genes and their regulatory networks of connections were in place, perhaps novel complex features could be developed mostly by random changes that accidentally form new signature sequences near various genes.[2] That would then activate the other gene whenever the master gene was turned on, perhaps leading to some new feature. The second group, styling their idea *facilitated variation*, emphasizes the ease of deploying an array of machinery to different locations, which, like the ectopic fly eyes, would generate a lot of variation much more easily than Darwin might have imagined.[3] Maybe that would give selection more to choose from.

If all that sounds distressingly vague, I'm afraid that is the gist of the argument. No one ventures a detailed, testable hypothesis about exactly how the original master genes and switches arose. No one actually spells out in anything like sufficient detail how—even after the first gene regulatory networks were in place—new switches and connections could be added at random to form complex novel features, let alone conducts experiments to show the proposal's viability. A very interesting question that might be asked, as an example, is how evo-devo manipulations might lead to the toothed gears in the legs of *Issus coeleoptratus*, discussed in Chapter 2. Exactly what master genes and which switches would change gradually to lead to that remarkable structure? Yet more than three decades after the discovery of master genes, no real progress has been made toward specifying in detail exactly how they could lead to an evolutionary explanation for some identifiable complex feature.

The unanticipated discovery of layers of control—master switches and the stunningly sophisticated genetic regulatory networks they activate—does not make the putative undirected development of life any easier to explain, as evo-devo enthusiasts seem to imagine. It makes it vastly harder. The need for a foreman and subcontractors to coordinate construction does not make it easier to explain how unintelligent processes could make a building out of bricks and wood and pipes and wiring. It shows it to be impossible.

Inclusive Inheritance

The neo-Darwinian synthesis began with the realization that discrete entities dubbed genes could control biological traits, such as the green or yellow color of the peas in Gregor Mendel's garden. It solidified in the mid-twentieth century around the seductive mathematics of theoretical population genetics. Later work identified genes as sequences of DNA, and Watson and Crick's discovery of the double helix showed the elegant way that genetic information could be passed down through the generations. But theoretical population genetics strikes some people as rather a bit too theoretical, and it seems that DNA is not the only substance that can be passed to offspring. Some EES proponents think that other inherited factors should receive much more emphasis than they have. They call their view *inclusive inheritance*.[4]

An additional way that information may be passed down to offspring is by "epigenetic" tags. During the lifetime of an organism DNA can be modified by a process called methylation (think of it as like adding diacritical marks to letters of the alphabet), and the modification can affect whether a gene is turned on or off. For example, a flowering plant called toadflax comes in two forms—one with the petals arranged in a circle and one with petals set off to the sides. The two differ in one particular gene that controls flower symmetry, but the difference is not in the nucleotide sequence of the genes, which are the same.[5] Rather, the radially symmetric variant gene is highly methylated. Mouse parents that have the same gene for fur color can have offspring with varying colors of fur if the gene has different degrees of methylation.[6] The point of these examples is that something other than the bare sequence of DNA—characteristics that can be acquired during the lifetime of the organism—affects the biological traits of the next generation.

DNA packaging may have a similar effect. In cells, DNA does not occur in splendid isolation. It's always associated with proteins, and the strength of the association can determine whether a gene is active. What's more, modification of the proteins can affect how strongly they stick to DNA. The more strongly they stick, the less likely the gene is to be turned on. Although it's not clear how such packaging could be transmitted across generations,[7] if nature found a way, then an offspring's inheritance would depend on more than just the DNA sequence of its parents. One recently investigated example involves the activity of the X chromosome of the tiny worm *Caenorhabditis elegans*.[8] Depending on the level of gene activity in the parent worm, the DNA-binding proteins are modified to a greater or lesser extent, and the proteins then go on to affect the level of gene expression in the baby worm.

Another way that a trait can be inherited depends on neither DNA nor protein, but on RNA.[9] A recent study showed that mice stressed early in their lives passed on their abnormal acquired behavior to their offspring. What's more, RNA isolated from stressed mice—when injected into fertilized eggs from unstressed mice—caused the abnormal behavior when they later grew up.[10]

Ever since the eighteenth-century French naturalist Jean-Baptiste Lamarck became a laughingstock for saying that giraffe necks lengthened over time because of the animals' striving to eat the leaves of trees, inheritance of acquired characteristics has been anathema to mainstream biology. The more science investigated, the more sense that negative judgment seemed to make. After all, during the formation of germ cells all identifying marks except the DNA sequence seemed to be stripped away, perhaps so that a new organism could start as a blank slate. But maybe nature has more tricks up her sleeve than was appreciated. Maybe Lamarck wasn't completely wrong.

It's not only molecular variations in cells that can potentially rival DNA's role as carrier of information between generations. The developmental geneticists Eva Jablonka and Marion Lamb have pointed to a handful of other possibilities.[11] For example, anatomy: a small mother may be constrained simply by her size to give birth to small daughters, whose offspring in turn will be small. Hormones too can have effects over the generations. Female gerbils who are litter mates of several males are exposed to higher levels of testosterone during gestation. That induces later behavioral changes, which lead the daughter gerbils themselves to have litters with multiple males, ensuring again that any females in the litter are exposed to extra testosterone. Antibodies transferred in mother's milk influence the development of the offspring's immune system. A mother's feces helps determine the bacterial content of a baby's gut. And a mother rat's caressing behavior can lead to hormonal changes and methylation of the offspring's DNA, which induces the same behavior when the offspring gives birth to the grandkids. Thus behavioral differences can be passed between generations independently of any sequence changes in DNA.

So inheritance of at least some traits can bypass DNA sequences. But how does that help evolution? In a nutshell, Lamb and Jablonka think the extra factors might increase the available variation for natural selection to choose from, at least for large animals. What's more, multiple genes can be modified by epigenetic modifications at the same time in the same animal, perhaps leading to a helpful combination of effects.[12] Yet just offering more variation more quickly is unlikely to help. The fundamental problem with inclusive inheritance is that there are so many different ways that a plant or animal or even bacterium can vary, almost all of them detrimental, that linking multiple necessary changes all at once is well-nigh impossible. As Richard Dawkins aptly pointed out in *The Blind Watchmaker*: "However many ways there may be of being alive, it is certain that there are vastly more ways of being dead, or rather not alive."[13] Charles Darwin's brilliance was to propose the accumulation of slow, tiny, step-by-step changes, which, his readership could imagine, might have the chance to integrate with each other and eventually lead to something useful. But the more simultaneous or even merely rapid changes one needs (as the theory suggests), the much more likely they'll be an incoherent jumble.

Niche Construction

Besides the genes and other molecules they transfer, parents can alter the physical environment in ways that make the success of their progeny more likely. This is usually listed by EES advocates under its own category, *niche construction*. One example goes all the way back to Darwin, who noticed that earthworms change the composition of the soil, in turn making the environment more favorable for their offspring. What's more, the descendants then adapt to the very environment that their ancestors modified. If the modern evolutionary synthesis holds that evolution is driven by the environment and if many organisms actively construct the environment in which they live, then, ask supporters of the extended evolutionary synthesis, is an organism driving *its own* evolution? It seems that creature and surroundings may not be nearly as independent as old-school thought assumed.

One has to be careful to draw a distinction between plants and animals simply altering their world and doing so in ways that would immediately benefit their successors—ways that could be called an inheritance.[14] For example, billions of years ago photosynthetic bacteria appeared that produced oxygen as a byproduct of their metabolism. At that point the gas was likely dangerous—molecular oxygen is a pretty reactive chemical, and modern cells go to great lengths to use it safely. Over a long time oxygen began to build up in the atmosphere and life started to use it efficiently to metabolize foodstuffs. Nonetheless, the initial oxygen excretion would likely not benefit the immediate or even near-term progeny of the original producers. Any eventual benefit was too far disconnected from the first appearance to count as a driver of evolution that would rival mutations in DNA.

Yet some effects of organisms on environments *are* immediate or near term and *do* directly affect the survival of the species. EES proponents list a number of them, but I think the best examples for illustrative purposes are the termite species studied by Scott Turner, of the State University of New York. One species, *Macrotermes michaelseni*, builds gigantic mounds up to 30 feet tall on the African plains. They are far from simple piles of dirt. Rather, the structure of the mound contains an elaborate tunnel system that allows fresh air in and heat out and keeps the level of oxygen remarkably constant.[15] In other words, the mound doubles as a "lung" for respiration and as a chimney to carry off heat from the breakdown of wood, most of which is generated not by the termites, but by the fungus they farm to metabolize indigestible cellulose into the sugar they do consume. The termites' environment is no simple hole in the ground. It's a home they make with much effort out of their physical and biological surroundings. Successive generations of termites must adapt to that home, so in a strong sense the organism *creates* the environment to which it adapts.

Another species of termite studied by Turner is *Microhodotermes viator*, which makes smaller mounds 3 to 6 feet in height. Once built by one generation, an abandoned mound can be recolonized by later generations of termites, if a winged reproductive male and female from another colony happen to land on it. If they do, they've hit the jackpot, because the design of the old mound captures scarce water more efficiently than does other land in the region. That makes it more likely the incipient colony will survive and continue to work the surrounding soil to improve its porosity. A later lucky mating pair can then inherit the further improved structure, and so on. Ultimately, the mound matures into a structure with a hardened bottom that can hold water above the water table.

Figure 5.1. A giraffe walks near a termite mound. The DNA of the giraffe stores much more information than does the structure of the mound.

Turner suggests that here the stereotypical roles of organism and environment are flipped: "It is not so much the termites that evolve to the prevailing arid environment; it is the environment that evolves to suit the physiology of the termites."[16] The key factor for genetic inheritance in his eyes is *longevity.*[17] Turner admits that genes last longer than most environmental structures, but thinks that a long-lasting, persisting environment could rival their claim to being carriers of hereditary information.

I think that's radically insufficient. The most critical characteristic of a would-be hereditary system is not longevity, although that's certainly necessary. Rather, it's the power to store large amounts of accessible *information* (Fig. 5.1). Even the simplest bacterial genomes contain hundreds of thousands of nucleotides that code for hundreds of sophisticated molecular machines that attend to the innumerable details of life, the lack of many of which is fatal. No matter how remarkable it is when compared to the surrounding environment, a termite mound is extremely simple when stacked up against even the most rudimentary of genetic systems. The hardened bottom that allows for collection of water in a *Microhodotermes viator* termite mound is a crude child's toy compared to any of the bacterial propulsion systems described in Chapter 2.

It is the living, gene-based termites that sculpt the mounds into their functional shapes. Although the termites certainly have to be adapted to the structure they form, there is little reason to think, and no evidence to show, that the mounds play a quantitatively significant role in the process. From a biochemical point of view, the same goes for niche construction in general. Many organisms do actively shape their environments in striking ways, but there is no reason to suppose that the environment does very much shaping in return.

Developmental Plasticity

The final broad area of the extended evolutionary synthesis is called *developmental* (or *phenotypic*) *plasticity*, defined as "the capacity of an organism to change its phenotype in response to the environment."[18] In other words, the shape, appearance, or behavior of a plant or animal can depend strongly on factors outside of itself. For example, the theoretical biologist Mary Jane West-Eberhard points to the marsh plant *Sagittaria sagittifolia*, whose leaves have different shapes depending on whether they develop under or above water.[19] The queen, workers, and soldiers of a termite colony have noticeably different shapes, yet they all come from the same parents. Their developments differ depending on how they are raised by other members of the colony, and none of the classes can reproduce successfully without the others. Depending on their status in the group, males of a species of cichlid fish are either large and aggressive or smaller with the color and behavior patterns of a female. The jaws of some cichlid fish change depending on their food source.[20] The color of some insects that mimic leaves depends on the color of the leaves surrounding them.

Even normal development of animals under quite similar conditions varies substantially because, at least for some features, cells and tissues have considerable leeway in how they grow. Most nerves do not have predetermined paths through the body, but "explore" during growth in search of muscle cells to innervate. Veins seem to follow the growth of nerves or vice versa. Even large organs such as the stomach and heart can vary in shape from one person to another.[21] Many other examples could be cited. The point is that an organism or tissue has a surprising amount of flexibility in the way it develops—it's not rigidly determined solely by the content of its genes without regard to its environment.

Some biologists see developmental plasticity as likely to play a large role in evolution. The idea is that, in the course of adjusting to changed environmental factors, an organism alters its behavior or development or both in a way that helps it survive. If the conditions persist, then perhaps its offspring too will develop in the same way. Over time one or more mutations then come along in the lineage's genome to accommodate the altered development—that is, a genetic change allows the plant or animal to normally develop to match the altered environment rather than to adapt through an altered route. West-Eberhard sees this as reversing the usual view of evolution: mutations in genes don't lead evolutionary change; rather, they follow it.[22] Over time, the environmentally induced alternative development of an organism may get locked into place by mutations to genetic control regions in what she terms "genetic accommodation."

A proffered paradigm of developmental plasticity is called, I'm afraid, the "two-legged goat effect," defined as "the exaggeration and accommodation of phenotypic novelties via adaptive plasticity."[23]

It's named after an unfortunate billy goat born without front legs that, rather than lie around its whole life, learned to walk after a fashion on its hind legs. What's more, at death the goat's anatomy was seen to have been altered in significant ways to accommodate its unusual locomotion.[24] Some EES supporters hint that a similar effect may have been involved in the evolution of the upright walking of humans.[25]

So what is one to make of the two-legged-goat effect in particular and developmental plasticity in general as an evolutionary driver? Although fascinating, developmental plasticity speaks exclusively to the *current* biology of the organisms, not to any future evolutionary potential, which remains extremely speculative at best. What's more, as we'll see in future chapters, although organisms do undergo mutations that accommodate them to their environments, those mutations bring their own difficulties.

What's now called developmental plasticity used to be known just as "nature versus nurture." That is, since organisms interact with their environment, of course their development will depend to an extent on some factors outside themselves as well as on innumerable internal ones. As with niche construction, however, there's little reason to think the two sets of factors are remotely comparable in importance.

Natural Genetic Engineering

An approach to problems with neo-Darwinism that is reminiscent of EES but actually quite different has been advanced primarily by James A. Shapiro, of the University of Chicago. He was a friend of the late Barbara McClintock, a pioneering geneticist and first woman to win an unshared Nobel Prize in 1983 for her work on mobile genetic elements. McClintock ran into considerable resistance in the mid-twentieth century for her views that genes could be regulated—turned on and off. She was later roundly vindicated. Her experience with opposition to her then unorthodox ideas by an entrenched old guard impressed Shapiro with the need to treat received wisdom skeptically, including received wisdom about neo-Darwinism.

Shapiro's 2011 book *Evolution: A View from the 21st Century* carries appreciative blurbs from some illustrious scientists who, like McClintock, were or are iconoclasts, including Sidney Altman, a Nobel Prize winner who showed that RNA could act as a catalyst; Werner Arber, another Nobelist, who discovered a class of DNA-manipulating tools called restriction enzymes; the late Carl Woese, who used DNA sequencing to unveil a third domain of life, the Archaea; and the late Lynn Margulis, who first proposed the seemingly preposterous but now widely accepted idea that mitochondria—the "power plants" of eukaryotic cells—had once been free-living bacteria. That's a huge amount of intellectual firepower endorsing a frankly disaffected attitude toward neo-Darwinism.

Shapiro calls his approach *natural genetic engineering*. Darwin pointed to human pigeon breeders as an analogy for what nature might do; that is, natural selection picks and chooses between advantageous and disadvantageous traits the way a pigeon breeder chooses which birds to breed based on desired traits. Shapiro points to the manipulations that modern biologists perform in their labs as an analogy for the way nature operates; that is, natural genetic engineering performs operations similar to lab manipulations in working with genetic material. After all, a large number of the tools scientists use to manipulate DNA in the lab come from the cell itself. In their work on DNA for experimental purposes, scientists often cut specific fragments out of the genome using molecular scissors called restriction enzymes; to stitch pieces of DNA back together they use an enzyme called ligase; to copy DNA in the lab a natural polymerase is employed; and so forth. What's more, small autonomous scraps of DNA taken from nature, such as viruses and plasmids, are used by biologists as vehicles to introduce foreign DNA into a cell or to rearrange a cell's own DNA.

In brief, Shapiro asks, if we can use those tools to engineer DNA, then why can't the cell itself use them both to meet current challenges and to evolve over time? The fundamental difference between this and Darwin's theory is that the nineteenth-century naturalist viewed variation as arising quite randomly, whereas with the benefit of twenty-first century hindsight Shapiro sees the available variation upon which selection can act as strongly shaped, and perhaps even guided, by the machinery cells themselves use in their lives.

Shapiro's view of the cell is elevated. The genome is not only a repository for information, he points out, but a read-write system, which can be manipulated by the cell itself. What's more, like our own computers, cells have formatting to direct their information-processing machinery to the right places. These include repetitive sequences of DNA that can give it special structures, organization of genes into regions that are accessible or inaccessible depending on need, and the epigenetic chemical markers of DNA discussed earlier. Genetic programs and information can be reused and repurposed, including those that control development of animal forms and those that determine the structure of protein regions called domains. All those abilities are used during the lives of cells, and all are controlled by them.

The cell's uncanny deployment of genetic engineering tools leads Shapiro to view it as sentient—not that it is conscious in a human sense, but that it acts purposefully toward its environment. So perhaps the cell can also use its capacities purposefully to direct its own evolution. Unfortunately, to many neo-Darwinists such talk carries a whiff of heresy. As Shapiro writes, even the phrase "natural genetic engineering" itself makes some biologists balk, because "they believe it supports the Intelligent Design argument."[26] To try to reassure such scientists, Shapiro emphasizes that natural genetic engineering principles are empirically observable facts. He writes that their role in evolution is open to experimental testing. He regrets that the nineteenth-century debate between mechanism and vitalism (roughly, the debate over whether living things depend on principles beyond physics and chemistry) has outlived its usefulness. He points to modern sciences, such as cybernetics, that investigate goal-oriented functions.[27]

Alas, I doubt the gatekeepers will be mollified. Speaking as a bona fide intelligent-design advocate, I don't think Shapiro has in mind anything similar to my understanding of the issue. Rather, there is a long intellectual tradition stretching back to the ancient Greeks such as Aristotle that discerns teleology in nature itself—not necessarily in something beyond nature. The view comes not from prior philosophical commitments, but from empirical observations that nature (most especially biology) acts for purposes. As Shapiro demonstrates, modern cellular and molecular biology have confirmed those observations in spades. Unfortunately, from the beginning Darwinism has itself made a heavy commitment to a mechanistic philosophy of nature that, to say the least, takes a jaundiced view of the notions of teleology and intelligence and looks forward to the day when they are finally and conclusively exposed as will-o'-the-wisps. Prejudice like that has to be confronted and fought directly, not appeased.

My own skepticism about natural genetic engineering has nothing to do with Shapiro's philosophy. Rather, as with neutral theory, the big problem I see is that it doesn't even try to explain the origin of purposeful systems—it takes them for granted. In order to even begin working, it requires sophisticated cellular tools to already be in hand. So where did the original, intricate, complex systems come from? Natural genetic engineering seems to have a big chicken-and-egg problem—it needs complex systems to make complex systems.

Even worse, as we will see in Part III, there is little evidence that the systems Shapiro cites are in any way creative beyond the boundaries of their current capacities. Laboratory and field evolution studies give no hint that, in the face of selective pressure, natural genetic systems engineer anything fundamentally new. Shapiro correctly notes that, "as many biologists have argued since the nineteenth century, random changes would overwhelmingly tend to degrade intricately organized systems rather than adapt them to new functions."[28] Yet the marvelous cellular systems he cites give every indication that they do the same thing when they move beyond their well-regulated limits.

Major Transitions

The increasing recognition of the failure of neo-Darwinism to account for life's functional complexities has made many biologists restless, creating an opening for novel ideas such as we've surveyed in this chapter and the last. Unsurprisingly, some Darwinian advocates have tried to compete with new arguments of their own. In the next section we'll look at a particularly popular one called game theory.

First, a bit of background. In the mid-1990s the eminent evolutionary biologist John Maynard Smith (who died in 2004) teamed up with Eörs Szathmáry, of Eötvös Loránd University in Budapest, to write *The Major Transitions in Evolution*. The eponymous events of the book were ones they judged to be particularly epochal in the history of life, not only because they were important in themselves, but also because they changed how the authors envisioned evolution could operate, either by affecting the way information is transmitted from generation to generation or by altering the division of labor among the components of life (Table 5.1).

The book's perceived importance can be measured by noting that the world's preeminent science journal, *Nature*, published an article by the authors summarizing its argument when it first came out,[29]

Table 5.1. The Major Transitions in Evolution (after Maynard Smith and Szathmáry, 1995)

The major unanswered question in evolution is: What do the arrows represent?		
Replicating molecules	→	Populations of molecules in compartments
Independent replicators	→	Chromosomes
RNA as gene and enzyme	→	DNA and protein (genetic code)
Prokaryotes	→	Eukaryotes
Asexual clones	→	Sexual populations
Protists	→	Animals, plants, fungi (cell differentiation)
Solitary individuals	→	Colonies (nonreproductive castes)
Primate societies	→	Human societies (language)

and the leading journal *Proceedings of the National Academy of Sciences USA* recently published a twenty-year update by Szathmáry.[30] In between those, a conference was organized specifically to discuss the book's ideas, and a follow-up book, *The Major Transitions in Evolution Revisited*, was issued in 2011 by the conferees. That's an extraordinary amount of attention, likely because the original book reflects how many Darwinians see evolution. So let's consider some of the points it argues that are germane to our focus here: whether the ideas *even try* to explain the elegant structures of life.

The authors introduced the ambitious book by frankly admitting the deep yearning behind it: "The real reason why we study [evolution] is that we are interested in origins. We want to know where we came from."[31] It begins with the hardest of topics, the origin of life. In the initial chapters, even before the first element ("replicating molecules") of the first transition in their scheme is reached, much hypothetical and extremely lucky chemistry must be invoked, from "chemotons" (theoretical reproducing metabolic units) to chemical evolution in clouds. Without resolving initial chemical problems, the authors nonetheless press forward, assuming that somehow

RNA-like molecules form with a mixture of sequences that could lead to another theoretical entity (called a "hypercycle"), which brings its own severe problems.

It's clear that, like everyone else before and after them, Maynard Smith and Szathmáry can't plausibly get even to the first of their major transitions. Nearly twenty years after the book was published, the respected origin-of-life researcher Steven Benner gave a bracingly honest but bleak assessment of the problem in an interview, rattling off a number of forbidding roadblocks that stand between simple chemicals and life (for example, "The first paradox is the tendency of organic matter to devolve and to give tar [that is, the tendency of organic matter to decompose into a viscous, oil-like mess]").[32]

And it doesn't get any easier even after essentially postulating the bare origin of supposed replicating molecules. Subsequent hypothesized steps leading to a first prokaryotic cell—the origin of proteins, the formation of a protocell, the origin of the genetic code, a supposed switch from RNA- to DNA-based heredity, and more—are just as intractable. An attendee at the later conference on the book noted that "prokaryotes are extraordinarily complex biochemical systems" and that speculation about their further evolution "presupposes the cell, and . . . the cell is itself the Mother of All Key Innovations."[33]

Investigating the problem that Eugene Koonin thinks takes $10^{1,018}$ universes to solve can be frustrating.

Once "extraordinarily complex" prokaryotic cells are conjured, the authors turn their attention to later transitions. Unfortunately, those explanations are also hopelessly vague, hardly more detailed than for the earlier chemical steps. For example, in one twenty-four-page chapter they consider how eukaryotes (cells with nuclei) may have arisen from prokaryotes (cells without nuclei), which is one of the most profound divisions in life. Three major steps of a possible scenario are listed, each of which is dispensed with in a mere two paragraphs.[34]

Although John Maynard Smith and Eörs Szathmáry's thinking on the major transitions of life is speculative, it's often speculative in a wonderfully revealing way. The general thrust of their approach to evolution after the origin of life can most clearly be seen when they consider symbiosis.[35] Two different kinds of organisms can live together in several possible kinds of symbiotic relationships: parasitism, in which one benefits from the association and the other is harmed; commensalism, in which one benefits and the other is not affected; and mutualism, in which both benefit. In their book the authors consider the evolutionary conditions under which one or the other of those relationships will prevail. They show that it depends on how a symbiont is transmitted to a host as well as on how each behaves toward the other. For example, if a symbiont helps a host, the host should allow it to grow, but if it hurts the host, the host should try to kill it.

Notice, however, that this is an *economics* approach: it considers only how an organism can act to benefit itself the most—that is, how to maximize evolutionary profits. No attempt is made to explain what is going on at the foundational molecular level. Rather, enormously sophisticated molecular machinery is treated as an insignificant detail that pretty much automatically appears when the need arises. For example, exactly *how* might a host kill an invader? The immune systems of both vertebrates and invertebrates are enormously complex. Yet the authors write simply that, when it is beneficial to do so, "the host evolv[es] defense mechanisms."[36] It seems that, like the bona fide economists we discussed in the first two chapters, Maynard Smith and Szathmáry don't concern themselves with the origin of biological widgets.

Game Theory

That economic approach isn't just a casual afterthought. During his long distinguished career, John Maynard Smith made his mark by applying *game theory* to evolution.[37] Game theory deals with the best strategies to win games (either frivolous or deadly serious ones) over the long term, where other players' moves affect your own. Among other areas, it has been applied to bluffing in card games like poker as well as to economic strategies for businesses or nations. Maynard Smith brought it to biology, where it has had enormous influence.

Maynard Smith used game theory to develop a concept called the *evolutionarily stable strategy*, or ESS (not to be confused with the EES, the extended evolutionary synthesis). An ESS is a pattern of behavior that, if adopted by a population of organisms, is stable against alternative strategies, such as the killing or tolerating of a symbiont under different conditions in the example above. Evaluation of an ESS can be useful in accounting for the behavior of organisms, but such economic approaches *presuppose* the equipment needed to accomplish any particular strategy. They *don't even try* to explain it. The fact that a host would be better off killing a parasite does not by itself call forth the machinery needed to do so, any more than the need for a nation to defend itself against invasion automatically supplies it with advanced weaponry. The biologist John Maynard Smith's economic approach to biology no more explains, say, the immune system of vertebrates than the economist John Maynard Keynes's economic ideas explain the existence of automatic coffee makers. To think otherwise badly confuses necessary conditions with sufficient ones.

After Maynard Smith's death, Eörs Szathmáry continued and built upon their work. In 2015 he and a colleague published a paper entitled "'Synergistic Selection': A Darwinian Frame for the Evolution of Complexity," in which they take aim at non-Darwinian explanations for complexity, including ones we have discussed, such as inclusive fitness, self-organization, and complexity theory à la Stuart Kauffman. The authors offered a novel Darwinian idea of their own.[38] Although they note that "Darwin never specifically addressed the evolution of complexity as such," they aim to rectify the situation: "Here we will describe an alternative approach that could be characterized as an *economic* theory of cooperation and complexity" (emphasis added). Like some aspects of the EES, the gist of the paper is that cooperation can help form complex systems.

Unfortunately, although the cost/benefit analysis (which the authors call "bioeconomics") they propose might tell you whether, say, a new kind of cell phone is economically feasible, it doesn't tell you how to build it. As valuable as they might be in describing necessary conditions for the success of some evolutionary strategies, economic theories are nowhere near sufficient to explain how complex molecular machinery arose.

Recall from Chapter 1 that, although the discipline of economics studies the trading of goods, it doesn't try to account for their existence. Nonetheless, even with that limited mission, specific economic conclusions are notoriously unreliable—unavoidably so, because the subject matter consists of very complex systems that depend sensitively on all manner of variables. Yet evolutionary biology promised to go beyond economics to explain not only the relatedness but also the origin of intricate biological machinery. Although nineteenth-century scientists such as Darwin could not have anticipated it, that mission was doomed from the start. The Principle of Comparative Difficulty allows us to quickly conclude that that exponentially more difficult task is impossible.

Darwin: Still the Keystone

The ideas discussed in this and the previous chapter are the most prominent of those advanced in the last few decades to augment, supplant, or otherwise rescue neo-Darwinism from its doldrums. Yet if you mentally line up any of them next to the descriptions of the complex biological systems in Chapter 2, the mismatch is stark. Neither the more mathematical ones we examined in the last chapter nor the more descriptive ones discussed in this one, neither neutral theory nor complexity theory, neither the ideas of the extended evolutionary synthesis nor the latest Darwinian innovations—none of them *even try* to account for the sophisticated machinery of life. None even try to account for the purposeful arrangement of parts.

Despite the carping of some EES proponents, other than Eugene Koonin (who relies on an infinity of universes), they all still leave the heavy lifting to orthodox neo-Darwinism, either explicitly or implicitly. By definition neutral theory can't account for functional complexity, so the job of explaining it is quietly passed to Darwin. At best evo-devo and other EES ideas kick in only after life has achieved an enormous degree of sophistication, which they tacitly assume to be provided by you know who. And since the two-legged-goat effect seems an even less likely candidate to explain the vertebrate eye than the theory it seeks to extend, that particular burden remains with the Sage of Down House.

Just who is rescuing whom here? Much as with teenagers who hold their dad in low esteem until they need some spending money, critics find that neo-Darwinism has its uses. Notwithstanding its own manifold difficulties, they all implicitly or explicitly defer to neo-Darwinism to account for the overwhelming majority of life's functional complexity. Thus neo-Darwinism is still the keystone of modern evolutionary thought, and the credibility of both are inextricably linked.

Yet, as we saw in Chapter 3, the theory is actually an amalgam of distinguishable ideas, including the often overlooked first and last theories. Unlike other components of Darwin's system, through long years those two crucial planks remained untested postulates. That has now changed. In the past few decades enormous advances in laboratory techniques have allowed experiments and studies that were heretofore impossible, studies that lay bare the largely unanticipated, startling effects of the Darwinian mechanism. The four chapters of Part III explore them.

PART III

Data

CHAPTER 6

The Family Line

For nearly a century after publication of the *Origin of Species* the study of evolution generally took two sorts of approaches: the description and classification of plants and animals (both modern and fossil) to construct a postulated tree of life; and the development of mathematical models of how hypothetical genes (whatever they were) should behave under hypothetical evolutionary conditions. Few experiments could be conducted at the time to rigorously test basic evolutionary ideas, because the needed tools were lacking. Inevitably, in the absence of touchstone experiments, theory and peer pressure came to shape evolutionary thinking. Although earlier biologists had been almost uniformly skeptical of Darwin's mechanism of natural selection, its easy mathematization helped propel it to dominance in the mid-twentieth century under the steady lobbying of the founders of the modern evolutionary synthesis.[1]

The heyday of Darwin's theory took place in the absence of answers to very basic biological questions such as, among others: What is the nature of a gene? Exactly how is genetic information physically passed to an offspring? Beginning in the 1940s with the discovery that DNA—not protein or something else—is the genetic material, experimental work progressively uncovered more and more basic facts of molecular biology that had been hidden from earlier scientists. For the rigorous study of evolution, however, it's not nearly enough just to have knowledge of current biological processes—one has to be able to determine which particular mutations have occurred in individual organisms and what their effect has been. Since mutations are molecular changes (alterations of the sequence of DNA and proteins), tools are needed to track the molecular level of life. Experimental tools to sequence DNA developed only very slowly in the 1960s and 1970s and then explosively in the 1990s and later.

Yet in order to stringently test Darwin's crucial first theory (the completely unguided randomness of variation) and last theory (repeated rounds of mutation and selection somehow form coherent complex functional features), it's still not enough just to track changes in DNA and their effects in a few plants or animals. One also has to examine huge numbers of organisms over many generations—or at the very least to examine the straightforward effects of mutations in modern populations whose history is well known. *Only in the past twenty years have such detailed, rigorous evolutionary studies even begun to be conducted.*

Compared to this recent work, all previous studies—as brilliant as the scientists directing them were, as reasonable as the hypotheses seemed in their day—were inconclusive at best and greatly misleading at worst. Now, by dint of terrific work by many biologists, made possible by stunning advances in laboratory techniques, we are in a much stronger position to judge Darwin's theories based on sound experiments, not on blinkered postulates. In this chapter we'll examine powerful new studies that—although clearly demonstrating the ability of Darwinian processes to account for small-scale adaptation—point decisively to strict limits on fundamental biological change by random mechanisms. Put more plainly, we'll see that Darwinian processes nicely account for changes at the species and genus levels of biological classification, but not for changes at the level of family or higher.

Darwin's Finches

Although Charles Darwin was absorbed by the puzzle of Galápagos mockingbirds, the Victorian naturalist is better known to modern readers for his work on the finch species that inhabit those islands. In his exploratory travels on the HMS *Beagle* Darwin collected thousands of specimens, including ones whose exact classifications were uncertain to him. Upon his return to England Darwin turned over bird specimens to a prominent ornithologist named John Gould for sorting. Gould pronounced that some were not the blackbirds or other kinds that Darwin had supposed, but new varieties of finches found nowhere else in the world. Gould divided them up into twelve species. Recent publications count fourteen, although, as discussed later in this chapter, that number is in flux.

Other people brought the birds to wide public notice. A 1947 book by the British ornithologist David Lack entitled *Darwin's Finches* gave the group its nickname. In 1994 the writer Jonathan Weiner popularized the work of Princeton University evolutionary biologists Peter and Rosemary Grant on the birds in the Pulitzer Prize–winning *The Beak of the Finch*. The Grants have written their own scholarly monographs, including *How and Why Species Multiply: The Radiation of Darwin's Finches*. And of course the birds are now featured in many basic biology textbooks.

Unlike other tropical birds, Darwin's finches are not brightly colored; their hues are mostly blacks, browns, and grays. They range in size from about 4 to 8 inches in length and .25 to 1.25 ounces in weight. The species are divided into four groups (genera) based on where they live, what they eat, or other characteristics: ground finches (six species), tree finches (five species), warblers (two species), and vegetarian finches (one species). In their struggle to survive, the birds eat whatever they can get in their beaks, including plant products (seeds, plant buds, nectar, fruit) and animals (insects, shellfish, eggs of seabirds). One subspecies, ominously named the "vampire finch," gets nourishment from drinking the blood of oddly tolerant seabirds, chiefly boobies. The strange behavior is thought to have started from the mutually beneficial practice of the finch picking pesky mosquitoes off the seabird. It goes to show that, like people, birds can take advantage of unwary trust.

The finch species differ from each other in more than just size—the shape and strength of their beaks vary as well (Fig. 6.1). Some have thin, probing beaks, which help the birds collect nectar; others have thick, stubby beaks, good for breaking open seeds. The bigger the beak and the bigger the bird, the better it can crack tougher seeds. Peter and Rosemary Grant have done unprecedented work to account for the differences. They have studied the finches for decades, taking yearly trips to the tropical islands to tag, measure, and observe the behavior of thousands of individual birds under all sorts of circumstances.

Although the Galápagos Islands are located in the open ocean, smack on the equator, the climate is surprisingly variable. On the up to mile-high volcanic structures there are many different niches: beaches; arid lowlands; intermediate climates; and cooler, wetter highlands. The amount of rainfall is one of the most important determinants of the quantity of vegetation that will grow and consequently of how many seeds are available for bird food. In the year 1977 a severe drought struck the archipelago; most plants withered, became dormant, and stopped producing seeds. The only seeds remaining tended to be larger and tougher, the better to survive the dry conditions. In turn, the finches that survived were also larger in body and beak, the better to crack open the remnant of seeds. Unfortunately, since most birds couldn't manage that, mortality was very high—85 percent of medium ground finches (*Geospiza fortis*) died that year.

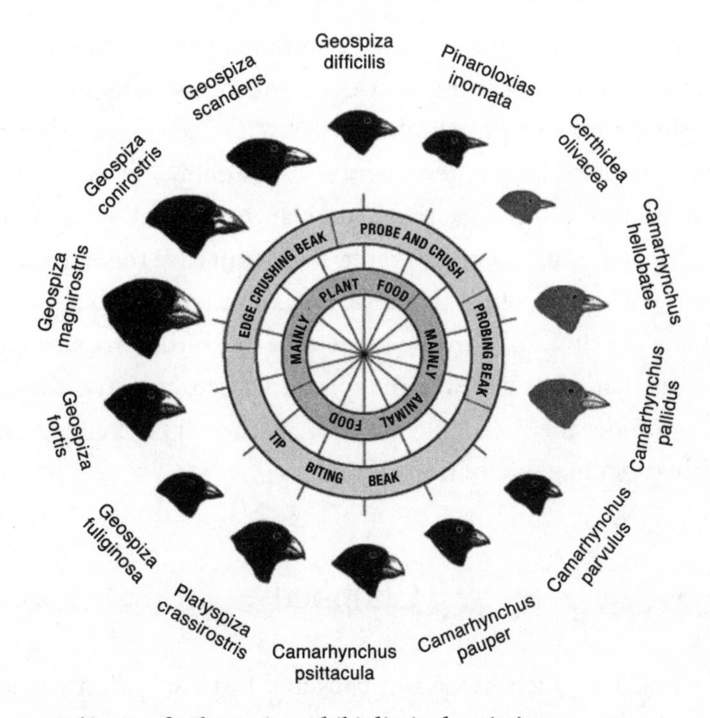

Figure 6.1. Galápagos finch species exhibit limited variation.

During and after the drought, the Grants carefully measured the size of the survivors and compared that to the size of earlier generations. They showed that on average the birds that weathered the drought were 5 percent larger, with proportionately larger beaks. What's more, their offspring were larger by about the same amount too. In other words, the Grants had meticulously documented evolution in the wild by variation, selection, and inheritance—a feat that had eluded Charles Darwin himself—and very likely the same sort of process that accounted for all the differences among the Galápagos finch species.

In subsequent years rains returned to the islands, and the researchers showed that the optimum balance for survival shifted with them. More plants grew that produce small seeds, smaller beaks were once again more efficient, and average size decreased. Over the ensuing decades body and beak size of the medium ground finch tracked changing climate conditions, varying by as much as 5 percent from the starting measurements of the early 1970s.

Weiner stresses in *The Beak of the Finch* that the notion that evolution is a hopelessly long-term process, invariably involving changes too small to see—an idea that even Darwin shared—is demonstrably false and, much to their credit, the Grants are the ones who falsified it with their frequent careful measurements. In fact, evolution happens continuously, in the sense that the birds that survive and the ones that don't have distinct physical characteristics that either help or hinder their struggle for life. And the very next generation is enriched in the genes of the winners.

Chopped

But what does such incessant, back-and-forth selection eventually yield? The Grants argue that "over the long term of many decades, centuries, or longer there could be a net trend towards a larger or smaller overall beak size."[2] Yet it's already clear that the long-term net trend doesn't go very far at all. The Galápagos Islands are the result of volcanic activity in the distant past, and the eponymous finches have been there for a very long time. It is estimated that thirty or so founding birds arrived there perhaps two million years ago and split into several species soon thereafter.[3] After a pause of several hundred thousand years, new species divisions came along sporadically since then.[4]

Thus the same variation, mindless selection, and relentless evolution of birds that the Grants have recorded so admirably for decades has actually been going on for about a million generations—*tens of thousands of times longer*—and involving a cumulative total of billions of birds. The results of the "net trend" produced by all that frenetic Darwinian evolution is a twofold variation in body length, shorter or longer beaks of greater or lesser depth (Fig. 6.1), and not much else. Beginning with something very much like a finch, Darwinian processes labored long and mightily in the Galápagos and brought forth . . . a finch. As John Gould informed Darwin a century and a half ago, all of the species remain recognizably finches.

As modest as the evolutionary results are on the surface, there is even less to the Galápagos finches than meets the eye. The Grants have been carefully reporting for decades that some of the finch species occasionally interbreed with each other to produce hybrid young and that in some cases the hybrids seem more vigorous than purebreds. In agreement with their observations, a recent report confirms that the large tree finch has disappeared from one of the islands, losing its status as a separate species by mating too frequently with another.[5] What's more, an even more recent publication argues that the supposed six separate species of ground finches are actually just one big species, with no statistically clear distinguishing traits among the populations.[6] Rather than fourteen total species on the islands, there may be only a handful. To describe what they think is going on, the authors coin the plaintive term "Sisyphean evolution." In the ancient Greek myth that is a paradigm of frustration, Sisyphus is condemned to eternally roll a boulder up a hill, only to watch it roll back down again as he approaches the crest. In the Galápagos Islands separate groups of ground finches apparently began to accumulate differences ages ago, only to repeatedly interbreed before the groups could divide into truly separate species. Even at such a limited level, Darwinian evolution has been frustrated for the better part of a million years.[7]

Jonathan Weiner was amazed by such changes as the Grants recorded, exclaiming that differences of 5 to 6 percent in finch beak size were the "difference between life and death."[8] But should we be surprised? Is a 5 percent difference tiny? To get a better handle on such changes, let's think in terms of the animal we're most familiar with—human beings. Suppose that in a diabolical experiment a Darwinian evolutionary biologist confined a group of a thousand adults to a tropical island. Unlike the Galápagos, this island was barren—no access to food or water except that provided by the researcher. In order to get to the food supply, the test subjects were forced individually to walk upright through a door. Hanging from the top of the doorway was a rapidly rotating ceiling fan with very sharp blades. The distance between the floor and the blades was 5.5 feet.

How much would a 5 percent difference in height make? An enormous one—the difference between life and death for many. Five percent of 5.5 feet is 3.3 inches. A fellow of average height, 5 foot 9, would have the top of his head chopped off. Someone 5 percent shorter, less than 5 foot 6 (like me), would walk through unscathed. *Of course* only shorter people would survive the hazard (many more women than men, just as many more male finches survived the Galápagos drought than females). And it also shouldn't surprise anyone with even the most basic knowledge of heredity that any offspring of the survivors would be shorter on average than the original group, since their parents were shorter. It's hard to imagine that the evil scientist's paper on the results would even be accepted for publication in a reputable journal, so self-evident are they.

Many other scenarios could of course be dreamed up to select for other traits besides shortness: putting the food on high shelves would select for height; making people walk a flimsy bridge over a chasm for their water would select against weight; and so on. The point of the gruesome examples is that such evolution as occurred in the Galápagos finches during the drought of 1977 is biologically trivial. It simply eliminates a segment of the population, and the remainder is left to reproduce. The remnant population has nothing at all that the starting population didn't have—nothing new. Indeed, it has *less* genetic variation than it started with. An average characteristic of the population—its size—taken as a whole has certainly changed, and one can call the process that caused the change "evolution" if one wishes. Yet for evolution to continue beyond the trivial, infusions of new variation—new mutations—that weren't present at the beginning must be added to a population.

Less Is More

Admirable as it is, the work of the Grants I've discussed so far doesn't really tell us how even limited evolution actually works, because we don't know what is going on within the finches that makes their bodies a little larger or smaller or their beaks a little thinner or thicker. Do those differences require brand-new genes? Or complicated rearrangements of old ones? Or maybe extra doses of preexisting "thick-beak genes"? Or what? Before just a few decades ago researchers would have been stuck at this point; even the smartest scientists would have had to rely on their imaginations to try to figure out what was going on.

No longer. With the dawn of the era of automated DNA sequencing techniques and clever new biochemical methods, altered molecular factors in mutant animals can be identified, the exact ordering of nucleotide letters in genes can be determined, and any changes from one generation to the next can be specified. The process isn't nearly as easy as that sounds—big experimental, theoretical, and conceptual speed bumps still abound. But in the hands of skilled workers, questions about the nature of the mutations behind evolution that had been impossible to address before can now be answered.

Molecular probes are now being used to investigate the evolution of Darwin's finches, and the Grants are leading the way. As mentioned in the previous chapter, it's been known for decades that some genes (by way of the proteins they encode) are "master controllers" that can switch on entire developmental pathways, such as PAX6 does with the eye (although most aren't as dramatic as that). The master genes aren't physically any different from other genes—they're still just stretches of DNA. But, like a start button on a computer or a key for a car, which can't do much by itself, they can activate a long already coordinated train of downstream events that yield impressive results. Realizing that some master genes that help control beak development might be altered between different species of Galápagos finches, way back in 2004 the Grants and some colleagues examined a number of known master genes and saw that one dubbed BMP4 was a lot more active in finch species with more powerful beaks.[9] A few years later their group showed that another gene for a protein called calmodulin was much more active in finch species with long pointed beaks.[10]

Yet even those results can't really pin down the evolutionary picture, because they don't show *why* the activity of those genes had increased. What *caused* the differences? What changes in DNA sequence—what mutations—were responsible? Not until 2015 did at least some of the answers roll in. Taking advantage of the (relatively) easy ability to sequence genomes these days, a team of scientists including the Grants determined the complete sequences of the entire genomes of 120 individual Galápagos finches and close relatives, including multiple representatives from each of the separate species and distinct populations of the archipelago.[11] That's over a *hundred billion* nucleotides of sequence!

As discussed in Chapter 4, most mutations are neutral and don't have any effect on survival, so identifying relevant mutations in such a large amount of information is, to say the least, a big needle-in-a-haystack affair. But with the aid of computers the Grants' team identified a half dozen master genes that were already known to affect head or beak development in mammals and birds and that also were discovered to have differences, somewhere in their vicinity, between blunt-beak and pointed-beak finches. (Interestingly, the previously identified important gene, BMP4, had no noticeable mutations in it, indicating that changes in a separate gene probably affect the activity of BMP4 itself.)

Most of the changes haven't been tracked down individually yet because they are outside of the regions that coded for the master proteins themselves and likely affect unidentified regulatory regions. The exception was for the gene that was statistically most strongly associated with differences between blunt-beak and pointed-beak finches—a gene called ALX1. In humans, the loss of ALX1 is known to cause severe birth defects due to maldevelopment of the head and face, which is exactly the kind of gene you'd expect to be involved in altering bird beak shape.[12]

ALX1 codes for a protein consisting of 326 amino-acid units whose sequence is very similar between classes of animals as various as mammals, fish, and birds. Of those 326 positions a grand total of 2—count 'em, 2—differ between the pointiest-beaked finch and the bluntest. That's like finding just one misspelling in each of the last two sentences of the previous paragraph, such as "strungly" instead of "strongly." The senses of the sentences are unaltered, just a little harder to read. At position 112 in the blunt-beak gene there's a P instead of an L; at position 208 there's a V instead of an I. All other positions are identical. The letters that occur in the blunt gene are the mutations, because they differ at those positions not only from the pointed-beak finch gene, but from the ALX1 genes of most other animals.

So what do those changes do to the protein? The authors write that "[computer] analysis classified both as damaging."[13]

Damaging. In other words, as in the case of the polar bear discussed in the first chapter, the mutations are predicted (based on computer modeling, not yet on actual experiments) to *impair* the normal function of the protein. How can mutations that damage protein function be positively selected in nature? Well, if the normal activity of the ALX1 protein during development helps make a beak sharper and more elongated, then hindering its activity could cause the beak to develop as less sharp and less elongated—in other words, shorter and blunter. If such beaks helped a finch survive a drought, the mutant gene would be selected. Thus a *beneficial* mutation can be one that *damages* molecular machinery. We will discuss this central idea extensively in the next chapter.

Sequence analysis shows that the two ALX1 protein mutations that differ between pointed and blunt genes did *not* first arise during the drought of 1977, when the Grants were observing smaller-beaked medium ground finches dying in droves. Rather, the mutations originated soon after finches arrived on the Galápagos, about two million years ago. Apparently, over vast ages, the same two very nearly identical gene variants—pointed and blunt versions of ALX1—have coexisted on the islands, shifting back and forth by selection and drift, first one gaining in frequency, then the other, helping populations of finches survive the vagaries of climate, world without end.[14] Just by chance, over all those generations on the islands, finches would also be expected to have been hatched that had mutations in all of the other positions of the ALX1 protein too. Yet, as we see from the results of modern sequencing, none of those other mutations were selected. Apparently none but the damaging ones could help those birds.

Drawing a Line

Carl von Linné—better known by his Latinized name, Carolus Linnaeus—was an eighteenth-century Swedish biologist who introduced the binomial ("two names") system of classification—the genus and species designations—so that people could more easily understand if they were talking about the same kind of organism. Because of his work a scientist can inform her colleagues that she's working with *Rattus rattus*, not *Rattus norvegicus*. Once you learn the terms, that's much easier and more precise than telling them, "I'm using that smaller black kind of rat, not, you know, that bigger brown kind." Pleased with his binomial work, Linnaeus defined hierarchical classification categories up to the most general levels of life then known—plants and animals. Over the years his classification scheme has been tweaked as microorganisms were added and some previously hidden intricacies of life came to light, but the basic framework is still pretty much the same.

The major divisions in modern biological classification are domain, kingdom, phylum, class, order, family, genus, and species. (There are also minor divisions, such as "subphylum" and "infraorder," but for simplicity we'll ignore them.) With each additional step of hierarchical classification in common, differences between two kinds of life decrease dramatically. For example, both flamingos and flesh-eating bacteria are living things; pelicans and pond scum are living things that are both in the eukaryote domain; larks and lice are eukaryotes in the animal kingdom; seagulls and sea squirts are animals in the chordate phylum; owls and ostriches are chordates in the bird class; crows and cardinals are birds of the order Passeriformes; black-headed tanagers and black-masked finches are Passeriformes in the family Thraupidae; Galápagos woodpecker finches and Galápagos mangrove finches are Thraupidae of the genus *Camarhynchus*.

As the work of the Grants shows so well, natural selection is relentless, brutally effective even in a single generation, seizing upon any variation however slight, cruelly separating the unfit from the fit. So for two million years the descendants of the original inhabitants of the Galápagos were constantly subjected to the most intense Darwinian selection. What amount of change in biological classification might have resulted from such searing selection? Some of the descendants might have differed from the ancestor in its domain of life or formed a new kingdom, or perhaps a novel phylum, class, or order. Yet none of those appeared. Instead, the descendants all remained even in the same family as the ancestor, differing only in the *two very lowest levels* of classification. The finch ancestors that first colonized the Galápagos Islands two million years ago are

Table 6.1. Classification of Galápagos Finches and Their Ancestor

Level	Ancestor	Descendant
Domain	Eukaryota	Eukaryota
Kingdom	Animalia	Animalia
Phylum	Chordata	Chordata
Class	Aves	Aves
Order	Passeriformes	Passeriformes
Family	Thraupidae	Thraupidae
Genus	Unknown	*Geospiza, Camarhynchus, Certhidea, Pinaroloxias*
Species	Unknown	Various

thought to have been related to tanagers, which are in the same family of birds that contain Darwin's finches.[15] Thus the ancestor and descendants would be classified as shown in Table 6.1.

One can think of the eight different steps of biological classification as like the different place values in an eight-digit number. To make it less abstract, think of the classification of a particular species in terms of a particular sum of money of hundreds of thousands of dollars, including change. The level of species is the pennies column; the level of genus, the dimes column; family, the dollars; and so on. Of course, two sums differ much less if they agree in all numbers greater than the pennies column (such as $132,547.38 versus $132,547.35) rather than in all the numbers greater than the thousand-dollars column ($344,217.19 versus $342,548.21). If classification is represented by an eight-digit number and the greater the difference in their numbers, the greater the difference between the biology of two organisms, what we find is that, after two million years of intense selection, after eons of the most fierce evolution, the finch ancestor and descendants differ in only the pennies and dimes columns—say from $213,754.36 to $213,754.83.

Why only such a tiny change? Is there something about their environment that would restrict the evolution of finches? That seems very unlikely. Not even counting invertebrates or life in the surrounding waters, the Galápagos Islands teem with all sorts of animals—tortoises, iguanas, mice, snakes—including many kinds of birds. There are boobies and mockingbirds and gulls and penguins and hawks and owls and more. Manifestly, plenty of niches exist that can support animal life very much different from finches. Yet millions of years of selection have left the finches very, very close to where they started.

Well, lengthy as it is, might two million years be insufficient for major evolutionary changes to take place? Demonstrably not. Most of the many, profoundly different animal phyla that arose during the Cambrian explosion did so in only about ten million years;[16] mammals diversified rapidly in roughly the same amount of time after the dinosaurs disappeared;[17] whales arose from a terrestrial ancestor in about the same time.[18] Surely we should expect at least one crummy new phylum, class, or order to be conjured by Darwin's vaunted mechanism in the time the finches have been on the Galápagos. But no, nothing. A surprising but compelling conclusion is that Darwin's mechanism has been wildly overrated—it is incapable of producing much biological change at all.

In *The Edge of Evolution* I argued that purposeful design was needed to account for life beginning from the very foundation of nature (such as the fine-tuned laws of the universe), through the elegant machinery of the cell, at least down to the biological level of *class* (birds are one class, fish are another). Without all of that basic stage setting, I wrote, life at that level could not exist. However, there was also then good evidence that the Darwinian mechanism of evolution by random mutation and natural selection could indeed account at least for the origin of new species, perhaps higher classifications. Somewhere between the levels of species and class, I argued, lay the rough boundary between what chance could account for in life and what required intelligent direction—the "edge" of undirected evolution.

With the progress of science in the subsequent years the uncertainty has decreased substantially, and it's time to update the estimate. From the wonderful research on Darwin's finches and other work I'll discuss in this book, it now seems reasonable to draw the line between the levels of family and genus. That is, chance plus selection can indeed give rise to both new species and new genera, just as Darwin envisioned, just as they did in the Galápagos. That's crucially important in enabling groups of organisms to diversify and fill disparate environmental niches. But, as a first approximation, Darwinian processes (or for that matter any other nonintelligently planned process) *cannot produce descendants that differ from their ancestor at the level of family or higher.* If one thinks of biological classification as an eight-digit number, then the depth of design in life can now be recognized as something like a hundredfold greater—reaching down two more levels of classification—than was possible to see just a decade ago. Put differently, fundamentally chance processes such as the Darwinian mechanism can affect only the pennies and dimes columns of life.

What variation can exist within a family? For the dog family, it's the difference between a domestic dog and a wolf and a fox. For the cat family, it's the difference between a lion and a leopard and a lynx. For the seal family, it's the difference between a ringed seal and a hooded seal and a bearded seal. That degree of variation can likely be achieved by random mutation and natural selection. What is the difference between members of two separate families? For birds, it's the difference between a swift and a hummingbird, or a woodpecker and a toucan, or a thrush and a starling. For mammals it's the difference between a cat and a dog, or a rat and a muskrat, or a porpoise and a narwhal. If my argument is correct, those differences required explicit design.

Clarifications and Caveats

To avoid confusion, it's critical to keep in mind that the concept of purposeful design is logically entirely separate from the idea of common descent—the idea that all organisms living today are descended from organisms that lived in the distant past. Some religious groups are opposed in principle to the idea of common descent. I am not. As I explained in earlier books, I think the evidence supporting descent is strong, and I have no reason to doubt it. Much more important than my own views, however, is that the concept of intelligent design contains no necessary opposition within itself to descent. As we saw in the Introduction, design is not a recent notion tied to sectarian beliefs—it's an ancient idea that can be traced back to the earliest pagan Greek philosophers.

Because design is not about common descent, the existence of fossils or even living organisms that seem intermediate between categories higher than family does not affect the argument. The design argument here is *not* that one higher category cannot descend from another through intermediates. Rather, the argument is that one higher category cannot descend from another *by means of an unplanned process* such as Darwin's mechanism. To cast doubt on design, then, it would be necessary for a critic to positively demonstrate that random, unguided processes could indeed lead to profound constructive biological change. Since that is precisely what proponents of Darwin's theory claim it can do, it seems only fair to ask them to demonstrate it to us skeptics, if they want to be taken seriously.

The conclusion of design challenges Darwin's *mechanism* of random mutation and natural selection—most especially his first and last theories; it doesn't dispute descent. As discussed in Chapter 3, the great majority of scientists accepted the notion of common descent soon after Darwin published the *Origin*, but were highly skeptical of his proffered mechanism of natural selection. The push for the neo-Darwinian synthesis in the mid-twentieth century herded most of the community into line but, as shown in the last two chapters, in the wake of stunning progress in biology many prominent modern scientists once again have grave reservations about what drives evolution. The argument of this book simply follows those dissenters' well-trodden path, with the important exception of arguing explicitly for the radioactive conclusion of the necessity for *real* teleology—*real* purposeful design—in the history of life.

An important limitation of the argument here—that design extends to the level of family—to keep in mind is that the system of biological classification is a human invention. Nature simply is; it cares not at all for our concerns. It is people who construct such things as classification systems to help them comprehend vast amounts of data. Since it is a human invention, it's easy for errors and ignorance to creep in. Both of those were present from the beginning with the Linnaean system, since bacteria—the most prevalent kind of organism on the planet—were initially left out of it. Even in our modern age, fundamental classification categories can be revised. In the 1970s the late University of Illinois biologist Carl Woese proposed that bacteria were not a single, uniform group, but were divided into two categories, the true bacteria and strange creatures then called archaebacteria. More recently, based on the sequencing of the genomes of a very large variety of birds, a group of researchers has proposed the existence of previously hidden fundamental relationships in that class.[19]

On the other, lower, end of the spectrum, it can be very difficult to decide not only whether an organism is a separate species, but also which genus or even which family to assign it to. As recounted earlier in the chapter, it's a matter of contention whether there are six separate species of Galápagos ground finches or just one. Darwin's finches are thought to be descended from tanagers, which had been placed in a separate family a few decades ago.[20] Now both groups have been placed in the family Thraupidae. Humans are placed in the family Hominidae with the great apes, but there are excellent reasons to suspect those differences are well beyond Darwinian processes. Giraffes are placed in the same family as okapis and shorter-necked extinct species, but one can easily wonder if random mutation and natural selection could account for their differences. Perhaps they should be placed in separate families.

Until recently classification had been based mainly on morphology and behavior. These days similarity in DNA sequences is the overriding criterion. As more DNA work is done, previous categories are sometimes found to be comprised of unrelated species, and older groupings are occasionally broken up, rearranged, and renamed. Another potentially serious problem is that identically named divisions between widely different categories might be incommensurate. A "family" in a plant order might be less or more complex than one in an animal order, let alone in microbial classifications.

Most of these difficulties stem from the fact that my setting of the current line between the levels of family and genus is not based on observing a purposeful arrangement of parts (which is the definition of design) in molecular data (which is the gold standard for deciding the strengths and limitations of natural selection). For example, in *The Edge of Evolution* I argued that design extended at least to the level of class, because different classes of vertebrates (fish, birds, mammals, etc.) have different numbers of basic cell types. Elegant work by the late Cal Tech geneticist Eric Davidson and others had shown that different cell types require different, exceedingly complex molecular genetic regulatory networks,[21] which I argued were beyond the capabilities of Darwinian processes to produce. In contrast, there is currently no good understanding of what molecular differences distinguish two kinds of organisms at the level of family.

Nonetheless, I think tentatively placing the line at family is well justified by the actual paltry results of the evolution of Darwin's finches over two million years, based on judicious application of the Principle of Comparative Difficulty first mentioned in Chapter 1. Recall that it says if a task that requires less effort is too difficult to achieve, then a task that requires more effort certainly is too. If relentless, mindless selection for millions of generations on an ideally geographically isolated organism that readily generates new species is nonetheless unable to approach the level of new families, then the establishment of new families or higher categories for organisms in less suitable circumstances certainly would be too. That principle will be strengthened below by a number of diverse examples of organisms that exhibit the same limited evolutionary behavior.

What's more, the conclusion is not without at least some molecular support. As I wrote above, the mutation that is most strongly associated with blunt beaks in the Galápagos is a damaging, degrading one, which has not changed in perhaps a million years. As we'll see in Chapter 7, a major discovery of recent evolutionary studies is that mutations that damage genes are sometimes beneficial. It's quite reasonable to think, then, that degradative mutations can help organisms to adapt and in the process can sometimes shift them into new minor categories of genus and species. At some level, however, new positive additional genetic information is needed to differentiate one category of organism from another, and family seems a strong candidate for that level. This argument, then, makes the prediction that significant positive genetic information will be found to be required to establish a new family classification. If so, then that will constitute a more rigorous, empirical criterion by which to mark the edge of evolution, one that resists the vagaries of artificial nonmolecular measures and is more easily applied across great biological differences.

Finches aren't the only example that shows Darwinian evolution does not work either at or above the level of family. In the next two sections we'll look at a group of fish called cichlids.

African Cichlids

Lake Victoria, on the eastern side of Africa where Uganda, Kenya, and Tanzania intersect, is the second largest freshwater lake in the world measured by surface area—over twenty-six thousand square miles, lagging behind only Lake Superior.[22] It is relatively shallow, with an average depth of just 130 feet, which gives it about the same proportions as a sheet of stationery paper. As climate changed over the ages, becoming wetter or dryer, the surface of the lake would rise and fall, even to the point of occasionally drying out. The last time Lake Victoria completely evaporated is estimated to be about seventeen thousand years ago, after which it subsequently refilled several thousand years later. Until the past few decades, during which disastrous wildlife-management decisions have decimated their numbers, the lake was host to about five hundred species of fish that are found nowhere else in the world. Since they are unique to Lake Victoria, and since there was no lake until relatively recently, that means those species must have evolved in place in just the past fifteen thousand years![23]

The newly evolved species belong to a larger group of fish called cichlids, which are well known to tropical fish hobbyists. Unlike Darwin's finches, cichlids come in many brilliant colors, from blood red to lustrous blues, greens, and yellows, and sport all manner of intricate color patterns. They range in length from just inches to 6 feet. Many species are collected by aquarium enthusiasts. Others, such as tilapia, are important food sources for humans, while still others are game fish for sport fishers.

Not only do various species of cichlids look different from each other; they behave differently as well, especially in feeding. Some specialize in eating other fish; others dine on snails or insects. Still others scrape algae from rocks or sift bottom detritus for whatever nourishment they can extract. Like humans, cichlids care for their young; both mother and father usually contribute to their upbringing. Unlike humans, some kinds of cichlids carry their offspring in their mouths for a while, the better to protect them from predators. Other cichlid species let fry swim on their own, but patrol the area to ward off danger as best they can.

Despite all that surface variation, the late University of California–Berkeley ichthyologist George Barlow remarked in his 2000 book *The Cichlid Fishes* that their basic body plan is quite conservative and that "what seem to be major changes in appearance have evolved with little alteration of the basic plan."[24]

Cichlids share a number of anatomical features. They typically have only one nostril on either side of their snout (other fish have two). Their lateral line, a dotted streak along their length that helps fish sense the flow of water, is interrupted—that is, it starts from the back of the head and proceeds about two-thirds of the way along the top part of the body. There it stops, but a second line begins directly below it and travels to the beginning of the tail. Unusually, the cichlid small intestine departs the stomach from the left side. Almost all modern fish have small bony structures in their ears, termed otoliths ("ear stones"), that help them sense motion. For unknown reasons, in cichlids the largest otolith, dubbed the saggita, is grooved.

Most especially, however, cichlids differ from other fish in the structure of their pharyngeal jaws. Unlike mammals, fish have two sets of jaws, one that helps to form their mouth and a second set in their throat, both of which have teeth. Some of the bones of cichlid pharyngeal jaws fuse together, and a protuberance on the lower jaw provides a point of attachment for more muscles, making the jaws more supple and powerful.

As a group, cichlids are neither young nor confined to Lake Victoria. They are found widely throughout Africa, Central and South America, Madagascar, and India. That distribution led some scientists to think that cichlids first arose when those lands were in contact with each other in the ancient supercontinent called Gondwanaland some 160 million years ago and were carried away when it split up into the modern continents. Recent DNA and fossil studies, however, give an estimate less than half that age, about 65 million years.[25] Thus cichlid distribution must be accounted for by other reasons.

In addition to Lake Victoria, cichlids dominate both of the other next largest African Great Lakes, Lake Malawi and Lake Tanganyika, each with about half the surface area of Victoria, although much deeper. Unlike Victoria, the other two are long thin "rift" lakes, filling gaps where the current continent of Africa is tearing apart. The rift lakes are much older than Lake Victoria's most recent refill—about a few million years for Lake Malawi and roughly ten million years for Lake Tanganyika. Despite their varying ages, the rift lakes both have about the same number of their own unique species of cichlids as Victoria, although there is much uncertainty about the numbers.

One Big Family

Cichlids from all regions of the world are members of the very same family, Cichlidae, part of the order Perciformes—perch-shaped fish. The cichlids from the three African Great Lakes are all placed in one subdivision of the family, the Pseudocrenilabrinae, which itself is estimated to have arisen in Africa about forty million years ago. Cichlids from other parts of the world have their own subdivisions. Because the most recent ancestor of the African lake cichlids is thought to have been a cichlid itself (probably similar to a species

Table 6.2. Classification of African Great Lake Cichlids and Their Ancestor

Level	Ancestor	Descendant	
Domain	Eukaryota	Eukaryota	
Kingdom	Animalia	Animalia	
Phylum	Chordata	Chordata	
Class	Actinopterygii	Actinopterygii	
Order	Perciformes	Perciformes	
Family	Cichlidae	Cichlidae	
------------	------------	------------	≈ The Family Line
Genus	Unknown	Various	
Species	Unknown	Various	

found in African rivers), as in the case of Darwin's finches the cichlid ancestor and descendants share six of eight major classification categories, as shown in Table 6.2.

The huge number of brand-new cichlid species in Lake Victoria has been widely hailed as the most spectacular example of evolution in (relatively) modern times and spoken of in breathless terms everywhere from popular articles, to student textbooks, to professional publications. Yet just as with Darwin's finches, if classification categories were represented as the eight digits of a sum of money totaling hundreds of thousands of dollars, cichlid evolution in the African Great Lakes would be confined to the pennies and dimes columns on the right. Compared to the vast sweep of life, that's just evolutionary change. Even the IRS tells taxpayers to round off the cents columns in their tax returns. If that were applied to evolutionary biology, all the touted variation of cichlids would be disregarded.

Recently a large group of researchers sequenced the entire genomes of a number of cichlid species from the African Great Lakes and related groups. Unlike for Darwin's finches, where contributing factors for the single conspicuous trait of beak shape could be tracked down in relatively discrete populations at the gene and protein level, the large number of varying traits makes that much more difficult for cichlids. Nonetheless, the gene for one mutant protein called EDNRB1 was reported.[26]

The mutant has changes in just a few amino-acid positions in two regions, one of which normally binds to a separate protein and the other of which is chemically modified after the normal protein is made. In other words, although it hasn't yet been explicitly investigated, two abilities possessed by the unaltered protein are likely lost by the mutant. This of course parallels the situation for Darwin's finches where mutations were calculated to be damaging to the master protein ALX1. Another parallel to the finches is that many variant genes in new cichlid species did not first arise when the cichlids diversified recently in Lake Victoria. Rather, they came from mutations that arose millions of years before then and were maintained throughout the ages in the ancestor species population.

Despite this pattern, I should add that not all selected mutations to diversified species in a family are necessarily degrading. For example, excellent work has shown that cichlid rhodopsin—a protein necessary for vision—has switched a single amino-acid residue back and forth multiple times between nearly identical functional forms that are a bit more sensitive to light at greater or lesser water depths.[27] Yet, damaging or not, any excursions from ancestor sequences are small potatoes.

It's interesting to note that the cichlids of Lake Victoria evolved in the last fifteen thousand years or so—the time that the current replenished lake has been in existence. Yet Lake Malawi is over a million years old, and Lake Tanganyika about 10 million. Despite the vast differences in age, all have roughly the same number of cichlid species. What's more, the independently evolved lineages of each lake often resemble each other closely, clearly demonstrating

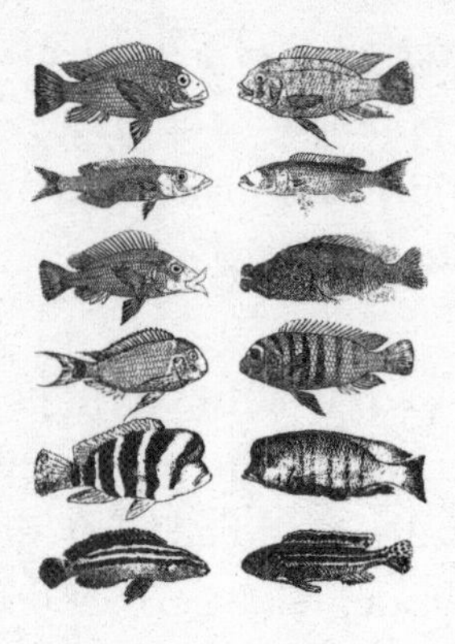

Figure 6.2. Cichlids of Lakes Tanganyika and Malawi. Fish species that evolved independently converged on similar forms.

the limited range of available variation, which apparently can appear very rapidly—and then just as quickly stagnate (Fig. 6.2).

It's also helpful to compare Darwin's finches to the cichlids. The two animals come from separate vertebrate classes—birds and fish—and occupy essentially opposite habitats, lakes or islands. They also have produced widely different numbers of descendant groups: together there are about fifteen hundred newly evolved cichlid species in the lakes—*a hundred times* the number of new finch species on the Galápagos. Yet, despite their enormous differences, both animal classes show the same severely restricted capacity to evolve away from their ancestors, never breaching the level of family.

The Same Old Story

Similar critical molecular studies have not yet been done for other groups as they have for Darwin's finches and the African cichlids, but other well-known luxuriantly evolved species fit the same classification pattern (Table 6.3). Like the Galápagos Islands, the Hawaiian Islands are an isolated archipelago, the result of volcanic activity over many millions of years, which have seen a number of colonizing groups evolve there.[28] One especially prolific group are relatives of *Drosophila melanogaster*, the common fruit fly, a favorite of genet-

Table 6.3. New Classifications Produced by Luxuriantly Evolving Groups

	Species	Genera	Families	Higher Classifications
Finches	14	4	0	0
Cichlids	~1500	~75	0	0
Anoles	~300	3	0	0
Honeycreepers	55	24	0	0
Fruit flies	~1000	2	0	0
Beetles	239	1	0	0
Silverswords	50	3	0	0
Lobelias	126	6	0	0

The Family Line

ics researchers. From apparently one ancestral species have evolved about a thousand modern ones, rivaling the number of cichlid species in African lakes. Yet the radiation has barely broken the level of genus, let alone family. Two thirds of the new fly species are in the same genus as the ancestor, *Drosophila*. Only one new genus, *Scaptomyza*, has evolved there.

Recently 74 new species of beetles were discovered on a Hawaiian volcano. All are members of the genus *Mecyclothorax*.[29] Including previously known ones, a total of 239 species in the genus are found on the island, all descendants of one colonizing event. A group of about 50 plant species dubbed the silversword alliance (named after a particularly flamboyant species) is endemic to Hawaii. The species are classified in just several genera, most of them in one genus. All are in the single family Asteraceae. The Hawaiian lobelias constitute 126 plant species in 6 genera. With their relatives in Africa and South America, all are part of the same family, Campanulaceae.[30]

A group of gorgeous birds, the Hawaiian honeycreepers, with widely varying beak shapes, all evolved on the islands. All are also members of a single family. (The honeycreepers had previously been classified in a separate family of their own, but molecular data now places them in a group with finches, Fringillidae.[31] Like the shuffling of Darwin's finches between two families, the Fringillidae and Thraupidae, this is a reminder of the vagaries of human biological classification systems.) In the Caribbean, anoles have diversified into hundreds of species occupying diverse niches on various islands.[32] Yet almost all are in the genus *Anolis*, with a few tentatively ascribed to several other genera. All are in the family Dactyloidae.

Whatever is causing these diverse kinds to spin off new species, it has precious little to do with what built the higher categories of life.

Animals of the large ancient island of Madagascar may be a bit harder to pin down. Endemic songbirds called vangas have diversified there into about fifteen genera of varying sizes and beak shapes, all classified in the same family, Vangidae.[33] Nonetheless, unlike the ancestors of Darwin's finches and other groups discussed above, the progenitor of the vangas is uncertain, presenting at least the possibility that it was in a different family.

More problematic are the lemurs, which are thought to have arrived in Madagascar fortuitously by rafting from Africa during a single colonization event over sixty million years ago and diversified into about fifteen genera in eight families—reaching all the way into the dollar column of eight-digit classification. The categorization of lemurs, however, is quite controversial, and they have no distinctive traits that aren't shared with other primates.[34] Thus the groupings may be an artifact of classification or, much more intriguingly, perhaps the result of the unfolding after colonization of intrinsic, intelligently provided information carried by the ancestor of lemurs, during a period when many new major categories of mammalian life arose. Firmer conclusions will have to await more extensive molecular analyses of the group.

Madagascar notwithstanding, drawing the limit of unguided, Darwinian evolution at the level of family seems quite compelling. Not only does it hold for birds (Darwin's finches, Hawaiian honeycreepers) and fish (cichlids), but also for reptiles (anoles), insects (fruit flies, beetles), and plants (silverswords, lobelias). These organisms of course represent widely divergent groups—differing from each other at the level of class (birds or fish or reptiles), through phylum (vertebrates or invertebrates), up to the classification level of kingdom (plants or animals)—that have very diverse modes of living, habitats, and population numbers. When such disparate starting points all lead to remarkably similar results, we can be confident that we have stumbled across a fundamental principle in operation.

And the fundamental principle seems very likely to be this: *minor random variations around a designed blueprint are possible and can be helpful, but are severely limited in scope.* For new basic designs such as those at the biological level of family and above, additional information is necessary, information that is beyond the ability of mindless processes to provide.

The overall results fit very well with comments by George Barlow:

> The reality of any classificatory scheme is often debated and is subject to continuous revision. A few taxonomists even want to do away with the binomial nomenclature, effectively getting rid of genera. A more radical proposition that is gaining followers is to do away with species names altogether.[35]

Species and genus classifications seem ephemeral likely because they are based on accidental attributes—on the caprice of random mutation and natural selection—which can arise through any number of serendipitous paths. Classifications at the level of family and beyond, on the other hand, are much more well grounded, because they very likely are based—directly or indirectly, consciously or unconsciously—on the apprehension of a purposeful arrangement of parts, that is, on aspects of the intentional design of the organism.

Darwin at the Molecular Level

Biology has come a very long way since Darwin first visited the Galápagos Islands, most especially including the key discovery that sophisticated machinery runs the cell. In the next three chapters we'll look at what random mutation and natural selection accomplish at the molecular foundation of life. The following chapter focuses on the damaging role of random mutation in Darwinian evolution.

CHAPTER 7

Poison-Pill Mutations

The elegant work described in Chapter 6 demonstrates that the very best examples of evolution in nature—multiple radiations of widely diverse organisms over millions of years that each produced from ten to a thousand new species—all bog down before they reach the classification level of family. As George Barlow remarked about the prolific African cichlids he studied, minor changes are plentiful, yet basic innovations are completely lacking.[1] The question of course is why. Why is unguided evolution so efficient at filling environmental niches with new species, yet gets stuck in the mud before any fundamental new biology is produced? Satisfying answers to that question were beyond reach just a few decades ago. But with the remarkable progress made possible by advances in technology, detailed understanding is close at hand, which will be the topic of this and the next two chapters.

Three factors interact synergistically to limit Darwinian processes. The factor I'll discuss last is actually the one I featured in *Darwin's Black Box—irreducible complexity.* Although irreducible complexity is most surely grasped at the molecular level I focused on, the severe problem for Darwin's theory of evolution in accounting for biological features that require multiple interacting components was one of the first major scientific arguments advanced against it (particularly by the noted biologist St. George Mivart in his 1871 book *On the Genesis of Species*). To many people over the years, both professional biologists and others, that's been an obvious difficulty, easily seen even in the larger features of plants and animals.

The other two factors, however, are more subtle and surprising, and a clearer understanding of their roles had to await the kind of large detailed molecular evolutionary studies that have only recently become available. Strangely, those two factors are none other than *random mutation* (which we discuss in this chapter) and *natural selection* (discussed in the next chapter), the essential components of the very engine of Darwinian evolution itself. In other words, as we'll see, *Darwinian evolution is self-limiting*—the same factors that make it work well on a small scale ensure that it doesn't go very far. Like a hot-air balloon, whose density allows it to rise in the thicker atmosphere near the surface of the earth, but also prevents it from climbing higher than the thinner air farther from the ground, random mutation and natural selection quickly adjust species to their environmental niches—and maroon them there. In this chapter we'll concentrate on the damaging role of random mutation.

The Most Definitive Evolution Experiment Ever

Everyone knows that the number of organisms you'd have to observe to get statistically significant evolutionary results is so enormous that no university animal facility or agricultural station could hold them all. And everyone knows that evolution takes a prohibitively long time—much longer than the average length of a research grant—so the process as a whole can't be followed directly. So everyone knows that the only way to study how evolution acts is to focus myopically on tiny parts of the elephant framed by

Darwin's theory, test them piecemeal, and then stitch them together—circularly using the theory to guide the interpretation.

Well, not everyone knows that, most especially not Michigan State microbiologist Richard Lenski. More than a quarter century ago Lenski had what seemed at the time to be a slightly wacky idea for a research project. He would grow liquid cultures of the common laboratory bacterium *E. coli* in a dozen separate flasks (for replicability's sake) in his lab overnight, just to see what happened. During that time the fast-growing bug went through about six to seven generations, until it had used up all the available food (dissolved sugar). Since bacteria are so small, the total number of cells in each flask the next day ranged in the hundreds of millions. In other words, the enormous number of organisms needed for statistically significant research results all fit in small containers on his lab bench.

Not stopping with just the one night's work, when he came in the next morning, Lenski withdrew 1 percent portions of each bacterial culture and used them to seed fresh flasks, which again grew overnight, producing another six or seven generations and hundreds of millions more cells. A tenacious man, Lenski—and a parade of grad students and postdocs under his direction—would repeat the ritual day in and day out, year after year, through the first Bush administration, the Clinton years, the two second Bush terms, and past the end of the Obama administration. As desktops gave way to tablets, as simple cell phones morphed into powerful information portals, as once-new cars became eligible for antique license plates, Lenski and his team persisted.

As I write, the lineage of bacteria at Michigan State has surpassed sixty-five thousand generations, which is equivalent to over a million years in the history of a large animal species like humans. So not only are there the big numbers of organisms from which to get real answers to evolutionary questions; there are more than enough generations for profound changes to occur too. Yet, as significant as those two factors are, equally important is that Lenski deliberately decided to let the cells simply grow, in the absence of artificial conditions or barriers, just to see how they changed over time. As he and some colleagues remarked confidently in reporting one study, "We did not artificially select cells on the basis of any phenotypic property. However, any mutation that conferred some competitive advantage in exploiting the experimental environment would have been favored by natural selection."[2] The bane of evolutionary studies known as investigator interference—where well-meaning researchers unintentionally poke and prod their study subjects in directions the investigators, but not necessarily nature, want them to go or focus their gaze on the portion of the elephant that theory tells them to concentrate on—all that was kept to a blessed minimum in Michigan. The cells were allowed to do whatever came naturally.

And what came naturally fit Darwin's theory perfectly. Although an imaginative movie producer might picture the cells evolving into some intelligent, slimy creature that stalked the campus at night, the actual results were more modest, but still exciting to academicians: the cells started to grow faster. Within a few hundred generations—about a month—of the start of the long-term evolution experiment, the descendant cells could regularly outgrow the ancestor cells in head-to-head competitions. (A huge advantage of working with bacteria is that Lenski could take portions of a culture at any time, which he did about every few hundred generations, store them in a freezer, and revive the samples at a later time to directly compare them in experiments with past or future generations. In a sense, in his freezer he has a complete living "fossil record" of the entire quarter-century experiment!)

In the first of his many papers on the work, Lenski measured the growth rate of the evolved cells and showed that after two thousand generations it was 37 percent greater than the ancestor's—a remarkable change.[3] He and coworkers also showed that there was some variation in the growth rates of the twelve replicate lines, as each apparently accumulated separate mutations, just as the neo-Darwinian concept of random mutation would predict. Interestingly, most of the improvement came in the earliest generations; the rate of betterment slowed with time. After fifty thousand generations, the most evolved cells grew only about 70 percent faster than the originals.

Very cool! So what exactly were those mutations that made the bugs grow faster? Were completely new genes made? Or old genes ramped up or rearranged somehow? Or what? Alas, in the early 1990s when the first work came out, it was well-nigh impossible to answer those questions. The genome of *E. coli* is millions of nucleotides long. Blindly fishing for mutations in an ocean of DNA was well beyond the contemporary technology. So Lenski's team pressed ahead diligently and cleverly to investigate the questions that could be answered at the time.

In subsequent papers they showed that the evolved bacteria had more descendants both because they grew faster and because they had a shortened "lag time" between cell divisions.[4] Oddly, the cells were fatter too—85 percent larger in volume. They demonstrated that increases in the growth rate of each cell line came in discrete waves as individual beneficial mutations (whatever they might be) arose and swept through the populations.[5] They showed that some cell lines grew better on alternative food sources, and others did worse.[6] To mimic sexual reproduction, they mixed the cell lines with other bacterial strains in hopes of increasing genetic diversity and the rate of adaptation; genes were swapped around, but no helpful evolutionary effect was found.[7] Ominously, one cell line turned into a "mutator," with a defective ability to repair its DNA, leading to a mutation rate more than a hundred times greater than normal.[8] Over the years another five of the starting twelve replicate cell lines would do the same.

It wasn't until the turn of the millennium that the first of the helpful mutations could be tracked down at the DNA level.[9] The watchful researchers noticed that all the evolved cell lines had lost the ability to metabolize a sugar called ribose. That clue gave them a target out of the great expanse of the bacterium's genome to zero in on. They sequenced the region around the multigene complex responsible for handling ribose in the ancestor and discovered that adjacent to it was a common viruslike "mobile element." The mobile element can be inserted and removed unpredictably by a cell's protein machinery at various points in the bacteria's DNA, often rearranging, inverting, or deleting chunks of DNA in the process. When they sequenced the evolved bacteria, they discovered that all twelve replicate cell lines had suffered massive deletions of the neighboring ribose genes, apparently facilitated by the mobile element. To put a point on it, a *beneficial* mutation (by itself that deletion mutation increased the cell's growth rate by 1 to 2 percent) turned out to be a *degradative* mutation, one in which the *loss* of a preexisting genetic capacity *improved* the bacteria's survival.[10]

How can that be? How can the loss of an ability be helpful? Well, what might be the quickest, easiest way to improve the gas mileage of your car, other considerations be damned? One way is to get rid of unneeded weight—toss out the spare tire, the hood, or even the doors or windshield. Of course, those things might be helpful in some future circumstances, but if the most important factor for your survival right now is the gas mileage, it would be beneficial to pitch whatever could be spared. If you were on a sinking ship and had to keep it afloat until it reached shore, throwing overboard any heavy unneeded equipment, no matter how sophisticated—computers, radios, cargo—is the winning survival strategy.

And it's not always just a matter of excess weight. Suppose in an emergency all traffic had to quickly evacuate a city, but traffic laws were still enforced (imagine all the cars were self-driving and were programmed to obey all laws). If the controls on a traffic light for an outbound route broke and it got stuck on green, allowing a large number of vehicles to legally pass through most quickly, many lives might be saved.

To switch back to bacteria, if an unneeded gene were active, breaking it would turn it off, saving energy. If a gene that would help make a useful product to outcompete other bugs were normally turned off, breaking the controls so the product would be made continuously would be beneficial. There are many circumstances in which getting rid of something can be helpful. And the more complicated and sophisticated a system, the more ways it can be broken in more situations to yield an advantage.

The Bottom Line

As the years passed at Michigan State, more and more mutations were tracked down. Other genes were quickly identified that also had been broken by mobile elements, also helpfully yielding faster-growing cells in the process, including genes involved in cell-wall synthesis, which may have allowed for the fatter cells mentioned above.[11] Lenski and coworkers used new technology to simultaneously probe all the genes of the mutant bacteria to measure which had higher or lower activity than the ancestor. Remarkably, they discovered fifty-nine genes that had changed their activity levels, either increased or (mostly) decreased them, all in the same direction in eight of the twelve mutant strains.[12] This was presented by some evolution popularizers as reflecting the repeated independent selection of multiple precise beneficial mutations.[13] In fact, as the authors directly stated in their paper, all those changes are due to the alteration of a single regulatory gene (dubbed spoT) for a protein that controls something called the "stringent response"—a process that normally signals other genes that already are attuned to it that there's an emergency due to the onset of starvation and to change their activity according to the preset plan.

The eight affected replicate strains all had point mutations in the spoT gene that caused single amino-acid changes in the encoded regulatory protein. Interestingly, all the mutations were different—that is, they had all changed the same gene, but at different places in it. Although the workers didn't explicitly test for it, that's the hallmark of a mutation that degrades or eliminates the activity of the protein it alters. It's difficult for a mutation to improve the activity of a protein, because most work very well already. Any improvements, if any are possible, tend to be limited to one or a very few potential positions. But it's easy to break or degrade a protein, just like it's easy to break or degrade a computer, by damaging it in any of a number of places. Protein activity depends on the interactions of a large number of amino-acid residues. Changing any one of them would have an excellent chance of hobbling a protein, sometimes (counterintuitively) helping a cell survive in the process.

Throughout the next decade Lenski and his collaborators continued their groundbreaking work, publishing insightful studies on the evolving bacteria, including many studies examining abstruse questions on topics dear to the heart of Darwinian evolutionary population biologists. Only in the year 2016, taking full advantage of the flowering of DNA sequencing technology, did Lenski and a host of coauthors publish the authoritative account of the results of fifty thousand generations of evolution of *E. coli*.[14] They sequenced the entire genomes of two representative cultures of each of the twelve replicate strains of bacteria after 500, 1000, 1500, 2000, 5000, 10,000, 15,000, 20,000, 30,000, 40,000, and 50,000 generations, for a total of 264 complete genomes! This yields an even clearer picture of their evolution than was provided by the sequencing of the genomes of the 120 Galápagos finches by Peter and Rosemary Grant and collaborators, discussed in Chapter 6, because Lenski's team probed many separate generations and because the genome of the bacterium is less than .5 percent of the size of the bird genome, so important changes can be identified with much more confidence.

The landmark paper contains two tables of the genes that were most frequently found to be mutated and thus very likely to have been the most highly selected—that is, to be the most beneficial. One table lists fifteen genes that acquired point mutations that changed single amino acids in the proteins they code for. Across the replicated cell lines all the genes had multiple mutations at different places—just as in the case of spoT, a strong signature of mutations that are likely to be degrading the activity of the protein. The other table is of sixteen genes that, like the ribose genes that were the first mutants to be identified, suffered repeated deletions or insertions of extra DNA, which usually kills a gene outright.

The bottom line is this. After fifty thousand generations of the most detailed, definitive evolution experiment ever conducted, after so much improvement of the growth rate that descendant cells leave revived ancestors in the dust, after relentless mutation and selection, it's very likely that all of the identified beneficial mutations worked by degrading or outright breaking the respective ancestor genes.[15] And the havoc wreaked by random mutation had been frozen in place by natural selection.

Making Distinctions

A few years ago I reviewed research done on laboratory evolution of microbes (including the work of the Michigan State lab done to that point) for a journal called the *Quarterly Review of Biology.* The article was titled "Experimental Evolution, Loss-of-Function Mutations, and 'The First Rule of Adaptive Evolution.'"[16] The goal was to reexamine lab evolution work from the past four decades and evaluate it in terms of the mode in which the microbes evolved. As Darwin himself knew, there are three very general ways in which an organism can adapt: (1) it can gain a new ability; (2) it can lose an old one; or (3) it can tweak or modify something it already has.

Because evolution can proceed in any one of those three ways, it's a question of profound importance to science to determine which one tends to predominate.

But when we try to categorize evolutionary events, we quickly run into a problem. What superficially looks like a gain or loss can actually be the opposite at the molecular level—a level Darwin and his contemporaries knew nothing about. To bring badly needed clarity to evaluating mutations, I divided them into three categories depending on how any particular change affected what I termed a "Functional Coded elemenT" or "FCT" (pronounced "fact").[17] A FCT is a stretch of information-bearing sequence that encodes a defined feature in either DNA or protein. Examples of FCTs are genes, control regions, protein-binding sites, protein-modification sites, and other such features. A given mutation, then, can either make a new FCT (which I dubbed a *gain-of-FCT* mutation), destroy an old one (*loss-of-FCT* mutation), or do something else—either tweak an old FCT in a way that leaves it still working or affect some noncoded feature of a cell (which I called a *modification-of-function* mutation).[18] Some mutations can be ambiguous and hard to classify, but most are straightforward.

Here's one hypothetical illustration. Suppose a bacterium becomes resistant to an antibiotic. At first blush, especially to somebody who gets infected and can't be cured by the now useless drug, that seems to be a significant gain of an ability by the microbe. But at the molecular level any number of events might lead to the same result, such as: (1) the bug acquires a brand-new gene that allows it to neutralize the antibiotic, which would be categorized as a molecular gain-of-FCT; (2) a control region that normally turned off a bacterial gene that could already inactivate the drug to a small extent breaks, allowing the gene to be active all the time, which would actually be a molecular loss-of-FCT (because the control region no longer works); or (3) a single amino-acid residue of a bacterial protein that interacts with the drug, leading to the death of the bug, is substituted by another kind of amino acid that blocks the drug from working; that would be classified as a modification-of-function. All three of these scenarios would yield a drug-resistant microbe, but the three reflect very different events at the molecular level.

Here's a second real-life illustration. Suppose a native of northern Europe traveled to central Africa and contracted malaria. From her sickbed she would likely gaze on the local inhabitants with envy at their resistance to the killer disease. She would certainly count it as a great additional ability compared to what her own biological makeup can do. Yet, since the molecular level of life has become clearer in the past sixty years, we now realize that the situation is more complicated. It turns out that different populations of people can be resistant to malaria due to different molecular mechanisms.

The most well-known antimalaria mutation is the sickle-cell gene, in which just one amino-acid residue out of hundreds in hemoglobin has been changed. The change causes the many millions of hemoglobin molecules in each of a person's red blood cells to be able to stick to each other in a specific way when the cell gives up the oxygen it's carrying in the capillaries. For some still unknown reason, that inhibits the growth of the malaria microbe (which lives in—and eats—a person's red blood cells), saving the lucky mutant person from sickness. In the *QRB* review I noted that, at the molecular level, that change is classified as a molecular gain-of-FCT—the gain of the ability of hemoglobin to specifically adhere to itself.

Other antimalaria mutations, however, don't confer any obvious new abilities at the molecular level. One (called hemoglobin C) changes the very same position as is changed in sickle hemoglobin, but to a different kind of amino-acid residue. In that case, the mutant hemoglobin doesn't aggregate, but still protects the person from malaria through an unknown mechanism. Since a FCT is neither gained nor lost in this instance, I categorized that as simply a tweak, or modification-of-function.

Very many other antimalaria mutations *break* genes or control regions and so are loss-of-FCT mutations. Hundreds of separate mutations have been identified that devastate copies of the gene for either the alpha chain or the beta chain of hemoglobin, leading to a disease called thalassemia. Many others destroy genes for glucose-6-phosphate dehydrogenase or band 3 protein. Still others mess up the control regions for fetal hemoglobin or Duffy antigen. As malaria researchers have noted, in addition to their helpful effects the mutations have less benign consequences, most especially "the great legacy of debilitating, and sometimes lethal, inherited diseases that have been selected under [malaria's] impact in the past."[19]

Those debilitating inherited diseases are the direct result of degradative mutations being positively selected for resistance to malaria. The crucial point is this: in order to properly assess what random mutation can do, we must evaluate evolution *at the molecular level.* Basing our judgment only on superficial effects can badly mislead us.

Can a loss-of-FCT or degradative modification-of-function mutation be exactly reversed under altered environmental conditions where it would once again be beneficial? Theoretically it could happen, but the odds are very much against it, for the same reason that loss-of-FCT mutations are much more likely overall than gain-of-FCT ones. There is only one way to exactly reverse a particular mutation, yet potentially very many ways to ameliorate a previous loss-of-FCT by a new mutation in another gene.

A good example of that comes from Lenski's work.[20] As mentioned, six of twelve cell lineages developed into mutators, which means their mutation rate was much higher—150 times greater than normal.[21] Lenski and colleagues showed that was due to just a single extra nucleotide that had been inserted into a gene that normally makes a DNA repair protein, breaking it. Later the researchers noticed that in one of the mutator cell lines the mutation rate had decreased by half, to about 75 times higher than normal. When they tracked down the cause of the modest improvement, it turned out to be a damaging mutation to a different protein, one that normally helps the cell metabolize nucleotides. So the unhelpful effects of damaging one gene were partially offset by a mutation that damaged a different gene. That's what random mutation does.

The First Rule

Just as with human mutations selected for malaria resistance, the great majority of beneficial selected changes in laboratory microbe-evolution experiments over the years are either loss-of-FCT mutations or modification-of-function mutations—no matter that they lead to an increased growth rate or other salutary properties. It's been known for a long time that the great majority of mutations that have a measurable effect on a creature's welfare are harmful.[22] The amazing but in retrospect unsurprising fact established by the diligent work of many investigators in laboratory evolution over decades is that *the great majority of even beneficial positively selected mutations damage an organism's genetic information*—either degrading or outright destroying functional coded elements.

Why is that the case? The simple reason is that the targets for damaging mutations are just much more numerous than those for gain-of-FCT mutations, so they'll be hit much more frequently.[23] Suppose a beneficial effect could be obtained by breaking or degrading a gene (Fig. 7.1). A modestly sized gene might consist of upwards of a thousand nucleotides. The ways one could break such a juicy target are legion. If an additional nucleotide were inserted anywhere between any of those thousand, that would yield a "frame-shift" mutation, likely destroying the cell's ability to translate the gene's DNA information into the correct protein. If a single nucleotide were deleted anywhere, the same kind of frame-shift problem would result. Larger insertions or deletions would frequently do the

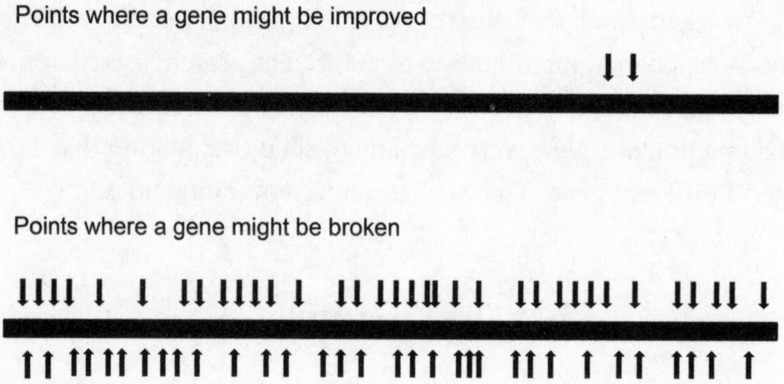

Points where a gene might be improved

Points where a gene might be broken

Figure 7.1. Mutations at many different points in a gene will break or damage it. Comparatively very few mutations might constructively improve a gene.

same. Also, because most proteins work very well already, point mutations that substituted one kind of amino-acid residue for another would very often degrade the protein's activity. In brief, just as it's easy to damage a car with a sledgehammer blow to any of a number of places, there are thousands upon thousands of ways to mess up a gene, so almost any random mutation to it would do the job.

Contrast this with a gene in which one of just a few nucleotides has to be mutated to yield a beneficial effect. That would almost always be the case for a new gain-of-FCT feature (such as a new protein-binding site or posttranslational modification site) because, just as in the case of a car, specific new functional features have to have specific structures in specific places. In other words, there are expected to be far fewer positions in a gene that can be changed to yield a helpful gain-of-FCT. Since mutations occur randomly, and since there are thousands of ways to break or degrade a gene, but perhaps just a handful to improve it constructively, the rate of appearance of a beneficial mutation that breaks or degrades a gene is expected to be hundreds to thousands of times faster than a beneficial mutation that has to change a specific nucleotide in a gene.

The gist of this rudimentary point can be succinctly stated as what I called the *First Rule of Adaptive Evolution* (in the epigraph at the beginning of the book it was slightly rephrased to make it more reader-friendly):

> *Break or blunt any functional coded element whose loss would yield a net fitness gain.*

It's called a "rule" in the sense of being a rule of thumb. It's not an ironclad law about what has to happen. Rather, other things being equal, it is what we should most often expect of random evolution in the great majority of cases. If throwing out a gene can help, if breaking a control region improves a species's lot, then random mutation will do it without a second's thought (after all, random mutation can't think).

Since the rule depends only on very general features of all living things—that is, on the structures of proteins, genes, and control regions and on the likelihood of breaking something rather than building something—it is expected to hold for all organisms from viruses, through bacteria, past single-cell eukaryotes, all the way up to the most complex animals, as it does in the examples we have discussed so far. It is called the "first" rule because the rate of mutations that diminish the function of genes is so much greater than the rate of appearance of a new feature. Thus damaging mutations will almost always occur first and so have the first opportunity, well before constructive mutations, to be positively selected if they are helpful.

Objections to the Rule

One objection to the rule might be that the laboratory evolution experiments on which it's based were artificial—they were conducted indoors, in flasks and Petri dishes, with far fewer organisms and for much shorter times than are available to nature. But the objection misses the mark by a mile. Yes, it is true pretty much by definition that lab experiments are artificial. Nonetheless, the same results are found in evolutionary events in nature, as we'll see later in this chapter and as illustrated by the effects on the human genome of exposure to malaria for ten thousand years. There also the great majority of adaptive mutations were loss-of-FCT or modification-of-function; only one was categorized as a gain-of-FCT, and that functional element (the protein-binding site that leads to sickle-cell disease) in itself is of distinctly dubious value. (Just as loss-of-FCT mutations can be helpful, gain-of-FCT mutations can be harmful.)

It is also true that even the longest, largest-scale experiments pale beside the resources of nature. Yet that perspective completely fails to grasp the significance of the insidious power revealed by small-scale work: it's not so much the *rarity* of constructive mutations that undermines Darwinian evolution—it's the *frequency* of damaging but helpful ones. Degradative but adaptive loss-of-FCT or modification-of-function mutations appear quickly even on *short* time scales, even in *small* populations. They don't need large numbers or long times to occur. Thus they will *always* be present *everywhere* in life much more quickly and in far greater numbers than constructive gain-of-FCT mutations. Damaging yet beneficial mutations will rapidly be selected when nothing else is available and compete fiercely with any gain-of-FCT mutations that might eventually arrive on the scene.

As relentless as the tide and as futile to try to resist, damaging yet helpful mutations will dominate unguided evolution over all time and population scales. As we'll see in the section after the next, even when the odd crude gain-of-FCT mutation sooner or later lumbers into view, helpful loss-of-FCT mutations will rapidly arrive to fine-tune the organism. They are inescapable.

A Fatal Implication

The almost oxymoronic "damaging but beneficial" mutations are the poison pills of Darwinian evolution. Plain old deleterious mutations aren't nearly as bad, because negative selection can weed them out. But degrading helpful ones are *spread* by *positive* selection. Even in limited cases where damaged genes are confined to just a segment of the population, the baleful effects stick around for a very long time, as malaria researchers have noted. If they become fixed in a species, however, the affected gene or control region is (barring very improbable reverse mutations) gone for good.

Burning down a gene or control region to help adapt to one demand means it is unavailable to help adapt to future ones. A fine illustration is *Yersinia pestis*, the bacterium that caused the Black Death in the fourteenth century. Analysis of its DNA shows that it's closely related to free-living *Yersinia* species in soil that either are benign or cause only mild digestive distress and are transmitted by contaminated biological waste.[24] About five thousand years ago *Y. pestis* apparently acquired two small DNA plasmids from other bacteria that carried several genes that allowed it to survive in fleas and so to be transmitted to people in a new way, by flea bite.[25] Those are classified as gain-of-FCT events.

The microbe then quickly adjusted to its new infectious lifestyle by *losing a hundred and fifty genes* that apparently were no longer needed in its new environment—which is of course a massive loss-of-FCT.[26] It seems quite safe to say that the bug is now stuck where it is, as an obligatory blood-borne pathogen. Although random mutation and natural selection might adjust it a bit further—perhaps allowing it to infect a different host at some future point—it will never be free-living again.[27] Along with the genes, it burned a lot of evolutionary bridges.

A fatal implication for Darwin's extravagant dream to explain life all by himself is this: the same primordial, inexorable, statistically inevitable process must do its work in all other organisms too, including the ones we discussed earlier such as Darwin's finches and the African cichlids. Their evolutionary radiations at the lowest classification levels and stasis at higher ones are flip sides of the same coin. Organisms quickly adjust to their environments by following the First Rule of Adaptive Evolution and are increasingly restricted because of it.

Citrus-Flavored Evolution

A final example of bacterial lab evolution drives home the relentlessness of degradative mutations. In 2008 Lenski's group first reported what was touted as a particularly impressive positive step in the evolution of his lab *E. coli.*[28] One morning, after more than thirty thousand generations of growth, a flask containing one of the twelve replicate strains seemed especially cloudy, indicating the presence of a lot more bacteria than usual.[29] Upon investigation they discovered that the bugs in the flask had developed the ability to eat citrate (a common cellular chemical found in abundance in citrus fruits) in the presence of oxygen, which normal *E. coli* can't do, although—and this is the kicker—the bug can readily eat citrate when oxygen is absent. For technical reasons the broth had contained a lot of citrate—a lot more than the sugar all the other bugs could eat. So with the extra food now available to it alone, the mutant quickly outgrew all others.

Further work uncovered the molecular basis for the new ability.[30] A protein that can import citrate into the cell has a control region next to its gene that switches it off when oxygen is around, which was the standard condition of the Michigan experiment. A mutation duplicated a chunk of bacterial DNA, serendipitously placing a different control region from a nearby gene next to the importer gene, allowing it to work when oxygen is present.

In my *QRB* classification scheme the mutation would be counted as modification-of-function—because no new functional coded element was gained or lost, just copied. Press reports, however, played up the new ability of the bacteria, and boosters of Darwinian evolution exclaimed that the alteration was a major improvement for the bug under its growth conditions. In fact, the authors of the study argued, since one traditional characteristic to classify a bacterium as *E. coli* is its inability to grow on citrate in the presence of oxygen, the mutant bug just might have taken a first giant step on the way to becoming a new species. (That speculation is actually quite modest. In thirty thousand generations Lake Victoria cichlids produced hundreds of new species. The bacterium has yet to produce one.)

But the stark lesson of this chapter by far overrides any squabbling about the significance of this or that particular mutation. To see why, consider the other mutations the citrate eater has suffered along its evolutionary journey. Like all of the culture lines, the citrate-using bacteria have lost the ability to metabolize ribose, suffered killing "mobile element" mutations to other genes, and fixed degradative point mutations in even more.[31] And, like five other replicate cell lines, the citrate user has turned into a mutator, with a greatly degraded ability to repair its DNA. Whatever the bug's fate from here, it has irrevocably lost the services of perhaps a dozen genes.[32]

But that's not all. In order to best accommodate the gene rearrangement that gave it the talent to eat citrate, several other mutations were found that fine-tuned its metabolism.[33] Even before the critical mutation occurred, a different mutation in a gene for a protein that makes citrate in *E. coli* degraded the protein's ability to bind another metabolite, abbreviated NADH, which normally helps regulate its activity. Another, later, mutation to the same gene decreased its activity by about 90 percent. Why were those mutations helpful? As the authors write, "When citrate is the sole carbon source, [computer analysis] predicts optimal growth when there is no flux through [the enzyme]. In fact, any [of that enzyme] activity is detrimental."[34] And if something is detrimental, random mutation will quickly get rid of it. Further computer analysis by the authors suggested that the citrate mutant would be even more efficient if two other metabolic pathways that were normally turned off were both switched on. They searched and discovered that two regulatory proteins that usually suppress those pathways had been degraded by point mutations; the traffic lights were now stuck on green.

Interesting as it is, the ambiguous citrate mutation that started the hoopla is a sideshow. The overwhelmingly important and almost completely unnoticed lesson is that genes are being degraded left and right, both when they directly benefit the bacteria and when they do so indirectly in support of another mutation. The occasional particularly noticeable modification-of-function or gain-of-FCT mutation can't turn back the tide of damaging and loss-of-FCT ones.

Invisible

Like water to a fish, sometimes the most important features of our surroundings escape our attention. In the history of evolutionary studies a lot of effort has been spent in understanding the behavior of beneficial mutations (almost always unspecified theoretical ones), somewhat less on deleterious ones. In the past fifty years the concept of neutral mutations has also been explored in depth. As a class, however, degradative and loss-of-FCT mutations are almost completely ignored. Certainly in their published experimental work researchers usually mention if a mutation involves the loss of some identifiable function. But it's treated as a bare fact, unrelated to any larger understanding of evolution.

It's not hard to understand why. Evolutionary theory affirmatively expects to be able to account for the development of all life on earth, which entails that the processes the theory posits built some very impressive biological systems. Of course any red-blooded researcher would want to show how that could occur. But since nature has been uncooperative, workers who persist in the field settle for building vague mathematical models or for conducting all-too-restricted experiments where the heavy hand of the investigator often unconsciously guides the results. Almost nobody studies degradation and loss-of-FCT for themselves, because theory implies they are at best peripheral to the main event and a distraction from the really important topics. To help disabuse ourselves of that view, let's look at recent evolutionary results that garnered some wider publicity.

In the only work I've seen until quite recently[35] that does focus on loss-of-function mutations as a general class, interesting in its own right, in 2013 researchers from Princeton and Columbia universities surveyed the literature and then conducted experiments of their own to see which bacterial genes could be broken causing the bug to grow better.[36] They showed that "at least one beneficial [loss-of-FCT] mutation was identified in all but five of the 144 conditions considered." In other words, *a bacterium could improve its lot by breaking a gene in over 96 percent of environmental circumstances* examined.[37] Several of the workers from the same group recently tested a more complex system, in which two different species of bacteria indirectly competed with each other, and showed that *E. coli* could adapt by damaging any of several genes.[38]

A brief comment on the original work by a news writer shows that the simple distinction between beneficial and constructive mutations has clicked for at least one person: "This study changes the widely held view that loss-of-function mutations are maladaptive."[39] Unfortunately the light seems not yet to have dawned on many others. News reports[40] in 2014 brought attention to a paper entitled

"Adaptive Gains Through Repeated Gene Loss: Parallel Evolution of Cyanogenesis Polymorphisms in the Genus *Trifolium* (Fabaceae)."[41] The gist of the research paper is that six different species of white clover that originally had the ability to give off poisonous cyanide (to discourage animals from grazing) all have variants that lost that ability by deleting the relevant gene. The news report frames the story as a question of whether evolution is repeatable. (The late Stephen Jay Gould famously said no.) The nose on its face—that evolutionary loss is always available to be selected—was ignored.

In 2013 a *Discover* magazine blog[42] highlighted a paper published in the journal *Nature* that tracked down the genetics of the ability of some horses to trot much more smoothly than others.[43] The trait is so popular with riders that horses with the heritable feature have been purposely bred far and wide by their primate keepers. It turns out that a single mutation in the gene for a nervous system protein is responsible. The mutation chops off one-third of the protein, degrading or destroying its function. The emphasis of both the publicity story and research paper is on how the trait is controlled by a single gene. Besides reporting the mutation, nothing is made of its being loss-of-FCT.

Horses and clover are nice, but what we really care about is humans—ourselves. A few years ago the *New York Times* ran a story about some people who seemed immune to developing adult diabetes. Like me, they were old and overweight, but otherwise relatively healthy. After screening many thousands of people in Sweden, Finland, and Iceland, researchers discovered a strong statistical association with a mutant copy of a gene for a protein dubbed ZnT8. What did the beneficial mutation do at the crucial molecular level? "The mutation *destroys* a gene used by pancreas cells where insulin is made" (emphasis added).[44] Another story the same year on the front page of the same paper told of a large study of people with a mutant gene named APOC3, who also had substantially lower cholesterol levels, shielding them from heart attacks: "The scientists found four mutations that *destroyed* the function of this gene" (emphasis added).[45] Unsurprisingly, both stories emphasized the medical angle.

I should point out that neither of the analyses above studied actual human evolution—they concerned only contemporary cases. Nonetheless, they are both fine illustrations of the benefits of breaking genes.[46] One case that does concern real, if rather humble, human evolution is that of a mutation in a gene involved with the production of earwax, thought to have arisen more than fifty thousand years ago.[47] In case you didn't know, earwax is categorized into two general types: wet (favored in warm climates) and dry (favored in cold climates). The mutation that results in dry earwax occurs in a gene dubbed ABCC11. It substitutes one amino-acid residue for another, which destroys the ability of the protein coded by the gene to work.[48] Whether it's diabetes, heart attacks, or the wrong kind of earwax, very often the quickest way for Darwinian evolution to mitigate a problem is to break something.

Next to people, we arguably are most interested in man's best friend, since we've apparently made so much effort to shape and select dogs over the centuries. In his review in the *New York Times* of my second book (he didn't like it), Richard Dawkins pointed to dog breeds as the premier example of the power of selection (albeit by humans, not nature) to shape animals as if they were so many lumps of plastic[49] (Fig. 7.2). But, at the DNA level, what exactly are the mutations behind the wide variety of dogs?

Largely degradative. Although they are very hard to track down, here are at least some of the known genetic changes:[50]

> Increased muscle mass in some breeds derives from degradation of a myostatin gene.[51]
>
> Yellow coat color is due to loss-of-FCT of melanocortin 1 receptor; black coat to deletion of a glycine residue from β-defensin.[52]

Figure 7.2. As cute as dogs are, much of the variation between breeds is due to *de*volution—to broken or degraded genes.

Coat "furnishings" such as long or curly fur come from mutations likely damaging to three separate genes.[53]

Six different genes control much of the variation in the size of dogs.[54] Half of them have likely degrading changes in the protein-coding region of the gene; the other three have tweaks in control regions that probably diminish the amount of protein made. All the mutant genes decrease the size of a dog.

Short muzzle is associated with mutations in the genes THBS2 and SMOC2, which probably lessen their activity,[55] and with a point mutation in BMP3 that likely damages the protein.[56]

White spotting results from small tweaks that decrease the activity of the MITF regulatory region.[57]

Short tails are associated with loss-of-FCT of the protein coded by a single copy of the mutated T gene.[58] Two copies of the mutated gene are lethal to a dog before birth.

Even the lovable friendliness of dogs toward humans (compared to rather less friendly wolves) is associated with the disruption of genes GTF2I and GTF2IRD1, whose destruction in humans leads to outgoing personalities plus mental disability.[59]

Dawkins is exactly right—dog breeding is a wonderful example of the power of selection acting on hidden random mutations. But now that we can investigate the molecular level of life, we see that the great majority of dog mutations unwittingly selected by us humans are very likely to be damaging, degrading, or outright loss-of-FCT ones.[60] Still, we shouldn't feel too embarrassed for our incompetence. If selection pressure in nature favored increased muscle mass of dogs, small size, or short legs, tails, or muzzles, there's no reason to suppose that some mythical blind watchmaker—acting on random mutations that it couldn't see either—would do any better.[61]

Speciation by Degradation

The examples just discussed show how common are beneficial degradative mutations, but don't link them directly to speciation. The example with which we started the book—the polar bear—does. Recall that the most highly selected genes in its pathway of descent from the brown bear were mostly damaged ones, including those involved in fur color and fat metabolism.[62]

That was reported in 2014. In 2015 an even more spectacular genome sequence was reported of a large, cold-climate mammal, but this time an extinct one—the woolly mammoth.[63] DNA recovered from two frozen fossils that died twenty thousand and sixty thousand years ago was compared to that of modern-day elephants. The charismatic species diverged perhaps seven million years ago—twenty times longer than did the polar bear and brown bear. Unlike the bears, which are species in the same genus, modern elephants and mammoths belong to separate genera in the family Elephantidae.

Many proteins will acquire one or two amino-acid changes over seven million years just by neutral drift—that is, by changes that do not affect their functions. Analysis showed, however, that of the approximately two thousand amino-acid residues found to be mutated in mammoths, about five hundred were likely to be damaging.[64] Another three hundred changes couldn't be decided, but a chunk of them too may be damaging. What's more, a further twenty-six genes were shown to be seriously degraded, many of which (as with the polar bear) were involved in fat metabolism, critical in the extremely cold environments that the mammoth roamed.

Although they haven't been directly tested, damaging changes to proteins that persist in a genome are likely to have been positively selected—that is, to be beneficial—otherwise they would tend to be eliminated by negative selection. Thus, although these are difficult matters to test directly, and although the more widely two species are separated in time, the harder it is to interpret changes, it seems very likely that degradative modification-of-function and loss-of-FCT mutations drove much of mammoth evolution. If so, then beneficial degrading changes explain not only modern evolution from bacteria to bears, but also the evolution of now extinct species that arose millions of years in the past.

In the *Origin of Species* Darwin argued that artificial selection—such as has produced various dog breeds—was an analogy for natural selection. He was more right than he knew: they both work predominantly by degrading genes. (As an aside, it seems reasonable to think that such a process may have a large, if indirect, role in extinction as well. The more genes that are degraded for short-term evolutionary adaptation, the fewer available for future adaptation, and the more brittle a species becomes. A further point is that the unexpected pattern of disparity preceding diversity seen in the fossil record—that is, new, higher categories of classification such as phylum and class preceding new, lower levels of classification such as order and family—comports much better with a mechanism of evolution by degradation of preexisting information than with a Darwinian mechanism, which predicts a pattern of diversity preceding disparity.)

The Dependent

It's easy to fall in love with evolution. Reading about a simple variant of a common moth that survives in a polluted environment better than its predecessor or bacteria that can eat industrial runoff can give one hope for the future, that maybe we humans can't mess up nature too much. Even stories about the galloping drug resistance of communicable diseases, although scary, can make you think about how nature maintains its balance. Newspaper reports of people who acquire immunity to the diabetes and heart disease brought on by our imprudent eating habits make us think that maybe we (or our children) can have our cake and our health too. New, pretty species of birds and fish evolving even in our own lifetime exemplify the fecundity of nature and, in Darwin's lyrical phrase, its endless forms most beautiful. He was surely right that there is grandeur in that view of life.

Yet it's incorrect. New life hasn't *e*volved. Overwhelmingly it has *de*volved—whether or not it strikes us as more attractive or impressive or useful than its forebears. Like an indolent scion of an old, wealthy family, life lives off its genetic patrimony—sometimes spending slowly, sometimes rapidly, but always taking in far less than it doles out. While it lasts, the fortune can shield species against the vagaries of the environment. But, as with polar bears and mammoths, the more it spends to adjust, the more restricted its options become. In extreme instances, such as with *Yersinia pestis* and probably some of the more misshapen dog breeds, a species completely runs through its legacy and is stranded in whatever biological niche it occupies last.[65] In any case, it will never have greater genetic wealth than what it inherited. That, at least, is the picture painted by the very best, most sophisticated evolutionary experiments the biological revolution has produced to date. And the principles revealed by the work are so fundamental that we must search for an even more basic principle to account for the source of life's wealth.

There are other ways to envision the unfolding of life than Darwin's, ways that are even more grand, ways that are much more congruent with contemporary scientific results, although perhaps not with contemporary scientific attitudes. We'll discuss those in the final chapter. In the next one, however, we'll probe another factor that makes Darwin's mechanism self-limiting—natural selection itself.

CHAPTER 8

Dollo's Timeless Law

Even as it helps a species to adapt to its present environment in strict conformity with Darwin's theory, random mutation is much more likely to damage genetic information than to build it. Over time that relentless tendency fences life in, making it less and less flexible. In retrospect, the easy production of new species and genera by widely diverse organisms—plants, insects, reptiles, fish, birds, as discussed in Chapter 6—coupled with the failure to generate any new higher classification categories are exactly what we should have expected from a blind process that can trade genetic inheritance for short-term gain.

In retrospect. Without the benefit of hindsight, however, even simple straightforward ideas can be surprisingly difficult to anticipate—including Darwin's theory itself. As "Darwin's bulldog," Thomas Huxley, is reported to have muttered after reading the *Origin of Species*, "How extremely stupid not to have thought of that." Yet as insightful as Darwin was, in a sense he had it easy. The main ingredients of his basic theory—variation, reproduction, inheritance, and selection—could all be seen on an everyday scale with the naked eye. It took imagination and acute powers of observation, yes, but little in the way of equipment.

On the other hand, the mysterious nature of heredity, although vital to a more accurate evaluation of his theory's scope, had to be put aside in Darwin's day and long thereafter as beyond the reach of the contemporary research technology. The very concept of hereditary information was then hazy at best, let alone that it would be encoded by a chemical substance in cells and the nature of the changes that the information could undergo were beyond imagining. Only with the very recent development of tools that reach to the molecular level of life with sufficient power to document individual genetic changes over generations has it even become possible to fully appreciate the long-term damage caused not by deleterious mutations, but by beneficial ones.

In the first part of this chapter, we'll see that sophisticated molecular research tools were also needed to show that it's not just random mutation that helps in the early stages of adaptation but then develops into a roadblock over time. Natural selection—the other half of Darwin's mechanism—does much the same, in two different ways. Then we'll turn to a puzzling question: If research clearly shows that the effects of natural selection and random mutation are limited, why do so many smart scientists still hold that Darwinism is the major force behind the development of life?

The Blind Metaphor

The primary way by which natural selection makes evolution self-limiting is by promoting poison-pill mutations. Whatever genetic alterations that help an organism survive and reproduce better than its competitors will be fodder for natural selection—even if the alterations make a species less able to adapt in the future. Over the generations they will sweep to fixation in a population. Of course random mutation is, well, random—it changes genetic material higgledy-piggledy, with no regard for the current welfare of an organism, let alone the future good of the species. Nonetheless, from Darwin's day onward many biologists attributed to natural selection the ability to finely sift mutations so that, somehow, over multiple steps a coherent basis for building an organism or complex biological feature would result.

That was always a bare hope, with little to no evidential support. Now that new laboratory methods are available to test that claim at the molecular level, we see that the hope is radically forlorn. Rather than guiding the construction of elegant biological machinery, selection predominantly scavenges a junkyard of broken or degraded parts. Degrading machinery can be useful for some purposes—perhaps because its function is unneeded at the time, and so the scrapped machine doesn't waste energy; or because in changed circumstances the product the machine made is now detrimental; or some other reason. But natural selection can't build a coherent new system any more than a pack rat can.

It is worth dwelling briefly here on the double-edged power of metaphors in science. Metaphors, of course, do not denote the thing they are applied to. They are vague analogies. It can be helpful in, say, a beginning physics class for an instructor to compare electrons passing through a wire to water flowing through a pipe or the Bohr model of an atom to planets orbiting the sun. But if metaphors are taken too seriously in science, they can be confusing at best; and they can often be actively misleading.

For Darwin's purposes, the metaphor of "natural selection" was brilliant. As nearly all of humanity thought until the mid-nineteenth century, life strongly appears to have been purposely designed. Thus the main problem for a fellow who wanted to account for life in the same ways that, say, physics or chemistry accounted for their subject matter was finding an unintelligent process that could plausibly mimic the action of a mind. The term "natural selection" helped immeasurably to persuade people that Darwin had hit upon a solution, because the metaphor hinted that the process itself possessed one of the most important abilities of mind—selection, that is, the *power of choice*. (The etymology of the word "intelligence" is "to choose between."[1]) If natural selection had the power of a mind, then maybe it really could explain life. In our day, another example is the "blind watchmaker"—Richard Dawkins's masterly metaphor for natural selection. That phrase also strongly insinuates intelligence.

Yet the metaphor of natural selection trades on an equivocation. The myriad processes involved in living and dying do not "choose" anything—they just happen. It can be loosely said both that intelligent breeders "select" the animals they want to reproduce and that an unintelligent sieve "selects" particles based on size. But breeders, of course, have a goal in mind, and they can follow up the first round of choosing with further rounds that are also directed to the desired end. A sieve has no such power. It will do the same thing over and over again. Thus there's no reason to think that nature can coherently follow one round of selection in further rounds. The ambiguous metaphor, however, greatly blurs the profound distinction.

For Darwin's mechanism of evolution, mutation is widely held to be the aspect that depends wholly on chance. On the other hand, natural selection is often called the *antithesis* of chance, as if it were akin to an architect planning a new building. The stated reason for its supposed power is that a feature that better adapts an organism to an environment has a much greater likelihood to increase in a population than otherwise.[2] Yet such a view depends completely, albeit usually unwittingly, on Darwin's last theory (discussed in Chapter 3)—the entirely unjustified assumption that repeated rounds of mutation and selection will add coherently to form complex systems. Since it was needed by Darwin's theory and couldn't be directly tested until recently, the idea was widely taken to be true. As we've seen, however, research now shows the premise is false. Human mutations that counter malaria aren't added up by natural selection to give anything coherent; they are all selected willy-nilly, regardless of whether they destroy a previously functioning system or not. Mutations that help *E. coli* grow faster at Michigan State aren't ordered to each other—whatever works at the moment is selected.

In hindsight, that is what we should have expected. Despite the boost in plausibility it receives from its metaphorical name, over *multiple rounds* natural selection is clearly nothing like the opposite of chance, no more than, say, gravity is the opposite of chance. Both of those phenomena are certainly directional, but only for *one* step. Beyond the first step there is *no* direction. Repeated tugs by gravity don't add up to anything constructive. A complex machine that repeatedly fell off a series of ledges would only break into smaller and smaller pieces. No one mistakes the results of gravitational attraction for that of a mind making a choice.

The same for natural selection. It will favor the increase in the number of organisms that do better in their environment for any reason, regardless of the basis for the variation. Selection is as unaware of whether a change in an organism helps in the present but hurts in the long run as, say, gravity is indifferent to the fate of a reckless competitive skier going much faster than other, cautious, contestants on a treacherous slope. No one should mistake the action of natural selection for that of a mind making a choice.

Specialized Tools

A second, slower, way in which natural selection makes evolution self-limiting has become apparent only in the past decade. Relentless selection will tend to fit already functioning molecular machinery more and more tightly to its present task, with no regard for future use. Eventually, like Gulliver restrained by the Lilliputians, it is immobilized by multiple weak evolutionary bonds, unable to be recruited even for relatively closely related tasks.

Before we dive into the science, here's a fanciful illustration of the problem on an everyday level. Suppose there were a plain metal rod that could be used for many sorts of tasks: as a crude fishing pole, as a crude baseball bat, as a crude hammer, and so on. As it happened, the bar was acquired by some people who wanted to build a house, so they used it for a hammer. As they used it, they modified it bit by bit for their purpose—first shortening it to make it easier to swing, then enlarging the end to better hit a nail, then putting rubber around the bottom to improve the grip, then lengthening the head perpendicular to the shaft—until finally it was the shape of an ordinary hammer one could find in a hardware store. After their house was finished, they decided to go fishing to relax.

But they couldn't, because they didn't have a pole (Fig. 8.1). By shaping the metal rod into a hammer they had made it less suitable for use as a crude fishing pole, no more fitted to that role than most anything else lying around the house. Shaping the rod more and more to a specific task made it less and less useful for other specific tasks. (Like anything else, however, it could still be used for nonspecific tasks that require no particular shape, such as being a paperweight or a doorstop.) Of course this is just a toy example, but the point remains: the more thoroughly adjusted to one chore a tool is, the harder it is to use it for another task.

What goes for simple tools, like hammers and fishing poles, goes in spades for more complicated mechanical ones, like can openers or mousetraps, let alone for electronic gadgets. And, as we'll see, it applies to natural selection working on exquisitely complicated biological machinery too. The very same role selection plays in Darwin's theory—adjusting a biological system to its current function—works to block the system from taking up a significantly different function. Like random mutation, natural selection limits Darwinian evolution on a large scale by promoting it on a small one.

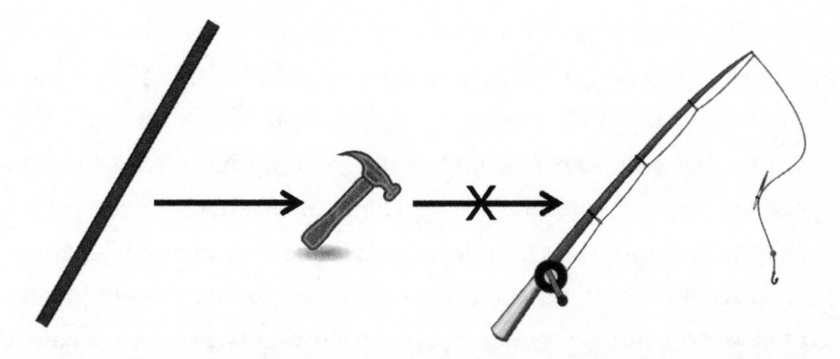

Figure 8.1. Cartoon of a simple metal rod "evolving" into a more specialized tool, the hammer, as described in the text. The new, complex shape hinders it from evolving into other specialized tools, such as a fishing rod.

Locks and Keys

The very best way to study evolution is simply to let a lot of organisms grow over many generations and then look to see what has happened at the molecular level, as Richard Lenski and others have done with microbes. A second best way is to investigate the changes that have occurred in populations with well-documented histories and connect them to mutations in their genomes, as Peter and Rosemary Grant have done with Galápagos finches.

Yet some questions can't be approached in those ways. To partially address some evolutionary questions, it can be helpful to reconstruct parts of life that are no longer extant and test how they change under various scenarios. In those cases we need to remain *acutely* aware that we're studying an artificial system and can easily be misled. Since such cases generally use molecular fragments of far larger systems, almost all of the actual biology—let alone many important evolutionary features—is left out. Conclusions drawn from such work are necessarily much more tentative. Nonetheless, they can be much better than nothing.

The Richard Lenski of this kind of approach is Joseph Thornton, previously professor of biology at the University of Oregon but now at the University of Chicago. Both Lenski and Thornton are terrific scientists and orthodox Darwinists whose work has placed that theory under an intense unprecedented light. Since about the turn of the millennium Thornton and colleagues have been studying how an ancient class of molecules called steroids (which are chemically related to cholesterol) interact with their receptor proteins. Proteins of course are long chains of amino-acid residues that—guided by precisely positioned electrostatic attractions between their constituent atoms—can automatically fold into different compact complex shapes, which are dictated by their amino-acid sequences. The complex shape of a protein allows it to perform its biological task, the way the shape of a hammer or a saw allows it to do its job. The task of steroid receptor proteins is to bind one or more members of the steroid class when present and signal the cell (by altering its shape) that it has hold of one. The cell then reacts in ways it has been set up to do, which we won't bother about here. The process can be likened to inserting a key (the steroid) into a padlock (the receptor) to open it. To work as needed, the shapes of the lock and key have to be closely complementary.

Thornton's group set out to investigate questions related to the evolution of two different kinds of steroid receptor proteins, the mineralocorticoid receptor (MR) and the glucocorticoid receptor (GR). Like two different sets of padlocks and keys, both proteins have very similar but slightly different shapes and bind steroids that also have very similar but slightly different shapes. The slight shape differences are crucial for determining which key opens which lock. Like all other of the many pairs of proteins whose shapes and amino-acid sequences are similar to each other, the genes for MR and GR are thought to have arisen when an ancestor receptor gene was duplicated in the distant past. Over time, the scenario goes, the initially identical genes accumulated different mutations, allowing their proteins to bind different steroids.

Thornton's approach was to compare the amino-acid sequences of both MR and GR proteins from many different kinds of modern vertebrates in order to infer the most likely sequence of the ancestor protein. Roughly, this involves finding a starting amino-acid sequence that could give rise to all the known modern sequences by way of the fewest number of changes. Using clever laboratory techniques his group then chemically synthesized the gene for the inferred ancestor protein, placed the gene in cells, and used the cells to manufacture the protein itself—a protein that likely hadn't existed on earth for hundreds of millions of years![3]

Analysis showed that the reconstructed ancient protein behaved very much like a modern MR receptor, binding the same several kinds of steroids (several similar "keys" fit this particular "lock") with about the same strength. So, in a question-begging sense, that takes care of explaining modern MR—its abilities were already present in the ancestor. But how about modern GR? How did it arise? By looking at the differences in sequences between modern MR and GR steroid-receptor proteins from many species, the investigators made educated guesses about which amino-acid changes might push the ancestor protein toward behaving more like GR.

Two possible switches seemed to be good candidates, so they used lab techniques to artificially make a protein with those changes. They also tested two intermediate proteins that had only one of each change (because random mutation would switch only one amino acid at a time). By itself, one of the two changes pretty much broke the receptor (that is, stopped it from binding steroids). By itself, the other change still allowed the protein to work as it had, yet it bound steroids much more weakly—about 1 percent as well as the original. With both changes together, binding was still very weak, but the relative strengths of the altered protein's binding to several kinds of steroids were a little different.

The upshot is that, although the work was technically very challenging and nicely done, what the results showed about evolution was quite modest indeed. The authors remarked that, since modern GR also binds steroids much more weakly than MR, the change was a step in the right evolutionary direction. Yet using a protein that already strongly binds several steroids as a starting point to design a slightly altered protein that binds the same steroids much more weakly—well, that's not exactly a ringing example of the fabled powers of Darwinian processes. Almost no biologist—certainly not myself—would be surprised by the results. But further results were soon forthcoming from the Oregon lab, this time ones that nobody expected, including me.

Dollo's Law

Louis Dollo was a nineteenth-century biologist who postulated that if a complex structure were lost in an evolutionary lineage (say, flight feathers in penguins), then it wouldn't reevolve there. Apparently the rule was proposed as a matter of convenience—he thought that if evolution could repeat itself, it would complicate the task of biological classification.[4] Over time his idea came to be known as Dollo's Law. Although Dollo's Law is taken with a grain of salt by most evolutionary biologists today for the case of larger biological structures, the question of the extent to which it applies at the molecular level is still debated.

In 2009 Joseph Thornton's group set out to explore whether something like Dollo's Law applied to steroid receptors.[5] Having shown in 2006 that the reconstructed ancestral steroid receptor could be changed by a few mutations into a weakened one they considered more similar to the modern GR receptor (as discussed above), Thornton's lab decided to investigate the reverse problem—whether, starting from the modern GR receptor (which binds just one kind of steroid), a pathway conducive to Darwinian evolution could be found back to the ancestral one (which binds several kinds, including the one bound by modern GR). As an analogy, if we show that a metal rod can be made into a hammer by a series of beneficial steps, can a hammer be turned back into a plain rod in the same manner?

After much impressive technically difficult work, their answer was *no*. The modern GR receptor is stuck where it is. It can't go home again, at least not with any reasonable probability by a Darwinian process. The reason is that the modern receptor has accumulated a number of other changes from the ancient one, some positively selected to help its function, others seemingly neutral.[6] Reversing them would be necessary to get the old function back, but changing them individually, one at a time, as Darwinism requires, either doesn't help or actively hurts, so natural selection would not be expected to favor them. The authors conclude that "the probability of all [necessary mutations occurring] in combination would be virtually zero."[7]

Although theirs is the first study with the necessary depth to address the question of the reversibility of protein molecular evolution, they are confident that the results will be quite general—that is, most proteins will be stuck in their present roles. In fact, they predict that further work "will support a molecular version of Dollo's Law." That is, "as evolution proceeds, shifts in protein structure-function relations become increasingly difficult to reverse."[8] If they are right, as there is every reason to think, the results throttle Darwinian evolution even further.[9]

Dollo Can't Tell Time

I remember my reaction on first reading the 2009 Thornton report—my jaw dropped. Although I'm a longtime skeptic of the grander claims for Darwinian evolution, I had always thought that it explained a good deal of biology. That's why I had found the group's

2006 paper so unremarkable. After all, the starting point was a protein that bound steroids and the ending point was a very similarly shaped protein that bound steroids more weakly. How hard could it be to switch one to the other? In the newer study they started with a protein that bound only one kind of steroid and tried to make a similar one that bound several kinds—in other words, one that was less specific. How hard could that be?

Like pretty much every other biologist, before reading the report my answer would have been: as easy as falling off a log. Even in previous disputes about the scope of Darwinian processes I would have conceded for the sake of argument that something like the ancestral steroid receptor could be turned into something like modern MR and GR receptors and back again by random mutation and selection as many times as a scenario proposed. But thanks to the great work of the Thornton lab, we no longer have to rely on our wildly inaccurate imaginations. We now know that's false.

The modern receptor could not give rise to a protein like the ancestral one by a Darwinian process, because the route is blocked by multiple small barriers that no one had any idea existed until now. It's hard to overstate the importance of the conclusion. As the authors write, it very likely applies to the great majority of proteins, which perform complex tasks by dint of their complex structures. The reason of course is that natural selection will fit all proteins—not just steroid receptors—to their current tasks without regard to whether a selected mutation hinders some potential alternative use or not. Neutral mutations will accumulate with the same utter disregard for distant utility. Drift plus selection will mire a protein in its functional place. (I hasten to emphasize that the conclusion is *not* that one kind of protein can't yield another, significantly different, functional one. Rather, the conclusion is that it is prohibitively unlikely to happen *by an unplanned process,* just as unaided nature would almost certainly not produce some of the dog breeds bred by intelligent humans.)

As politicians with something to hide often say during election campaigns, this isn't about the past; it's about the future. Who cares if proteins can't reevolve past functions? The pivotal implication of the work is that *future* changes of function by an unguided process would also have been severely impaired. That's because, like for many of the basic laws of physics, at the molecular level of evolution there's no telling past from future. For example, just as a movie of billiard balls bouncing around an ideal frictionless, pocketless pool table would look the same played backward as well as forward, a mutation in a protein substituting, say, a valine for a leucine would look the same as the reverse. Selection fits a protein as closely as it can to its *current* task, and to nothing else. Just as selection tends to block off reversion to a past state for steroid receptors, it would tend to block off future states for it and other proteins as well. Since no protein was ever without a prior history of selection, at pretty much all stages of life on earth all proteins would have faced the same hurdles to evolution.

The Oregon group's work on steroid receptors points strongly to a simplified justified twenty-first-century version of Louis Dollo's arbitrary nineteenth-century law. I'll call it Dollo's Timeless Law (Table 8.1). The original law looked only backward in time, ruling out for bare convenience the reappearance of any visible feature that had been lost in a lineage. In contrast, a time-independent, molecular-level, experimentally well-supported Dollo's Law essentially shuts off both the past *and* the future to Darwinian evolution. Not only is the *re*appearance of a complex functional molecular feature ruled out for all intents and purposes; so is its appearance in the first place.

The Necessity of Experimentation

The unexpected difficulties for the evolution of steroid receptors were only discovered because Thornton and his colleagues examined

Table 8.1. Dollo's Law Compared to Dollo's Timeless Law

Dollo's Law	Dollo's Timeless Law
Any evolutionary pathway from a *past* complex functional state of a protein to a significantly different *future* functional state of the same protein is unlikely to be *reversed* by random mutation and natural selection. The more the states differ, the much less likely that a *reversible* pathway exists.	Any evolutionary pathway from a . . . complex functional state of a protein to a significantly different . . . functional state of the same protein is unlikely to be *traversed* by random mutation and natural selection. The more the states differ, the much less likely that a *traversable* pathway exists.

them in unprecedented detail. If they hadn't done so, we would have been left with a very misleading view of the ease of their evolution. So it's good to remind ourselves at this point that even such excellent laboratory work gives only a very incomplete understanding of the likelihood of steroid-receptor evolution. Many, many other critical biological and evolutionary factors that would play a role have been left unexamined. At most, laboratory work gives a very-best-case scenario. The evolutionary prospects can't get any better than their work shows, but can easily get much worse. That's because the work examines difficulties for only a few steps of a much longer evolutionary pathway where many other unanticipated obstacles might well be lurking. If you had to trek blindfolded through a hundred miles of wilderness and the first mile was practically impassable, that would not lead you to think the rest of the journey would be smooth. Quite the opposite.

Darwinism fails when reasonably probable mutational routes to selectable structures are unavailable and whatever pathway was in fact traversed during the unfolding of life happened despite all odds. Even card-carrying Darwinists should agree that a massive role for serendipity in evolution drains most of the explanatory power from Darwin's mechanism and replaces it with a mere shrug of the shoulders—or with Eugene Koonin's *Twilight Zone* multiverse. In a 2014 paper Thornton shrugged, chalking up the extraordinary result to "historical contingency"—in other words, dumb luck. Consciously or not he was echoing the Nobelist molecular biologist Jacques Monod, who nearly fifty years ago wrote an influential book asserting that life and the universe were governed solely by chance and necessity.[10] But that's a false choice. We ourselves know those are not all that exist. There are also mind and plan. And for those willing to see, from the surface to the deepest levels, their effects are chiseled boldly into life.

Evolution by Gene Duplication Revisited

Rather than a fluke, there's every reason to think Thornton's research results showing the difficulty of evolving steroid receptors are typical.[11] That suddenly calls into question a huge chunk of what had been considered settled evolutionary knowledge. Almost all proteins whose origins scientists think they can explain are supposed to have been derived from ancient preexisting proteins (whose origins are entirely unexplained) by gene duplication plus accumulation of mutations—the same processes that gave rise to the modern MR and GR proteins from a single-steroid receptor gene in the past. The seminal book developing the idea, *Evolution by Gene Duplication*, by the late geneticist Susumu Ohno, dates back almost half a century. Up until now the lighthearted assumption was that the acquisition of significant new abilities by divergence of gene duplicates through random mutation and natural selection was unproblematic. Yet Thornton's work demonstrating the severe difficulties with even comparatively simple transformations—at least by a Darwinian mechanism—calls into grave question whether the others, even the most familiar, could have developed that way either.

As just one example, the alpha and beta chains of hemoglobin are universally thought to have arisen in the distant past from a myoglobin-like precursor by gene duplication and divergence. In fact, more than twenty years ago in *Darwin's Black Box* I myself pointed to them as examples of what Darwinian evolution could likely do.[12] They certainly derive from a common gene, but whether that could have happened by a Darwinian process—whether a comparatively simple oxygen-binding protein could without direction yield the sophisticated oxygen-delivery system that is hemoglobin—is now very much an open question.

What's more, glib stories in the evolutionary literature about how complex molecular machines comprised of many different proteins may have arisen as individual proteins and then come together into a coherent new system are now even more suspect than they were (if that's possible). The reason is that, like steroid receptors, the individual proteins would have been honed by natural selection to fit their ancient roles and so have been impeded from evolving new properties, including the ability to bind to and cooperate with other proteins in a larger complex (Fig. 8.2). The Principle of Comparative Difficulty tells us that if minor changes in a single protein are substantially blocked to Darwinian processes, major changes in many proteins certainly will be too. The more alterations that would have been required, the very much more unlikely it would have been for it to have happened without direction.

Everyone—including me—thought we knew a lot more than we did. Still, no one should now make the opposite mistake and leap to the conclusion that no development of protein function at all can occur by a classical Darwinian mechanism. As I mentioned in Chapter 6, a cichlid rhodopsin has apparently switched multiple times between two forms sensitive to different wavelengths of light,[13] and a recent study of Andean wrens discovered a point mutation that caused its hemoglobin to bind oxygen more strongly.[14] Those and similar simple examples are straightforward. However, whenever

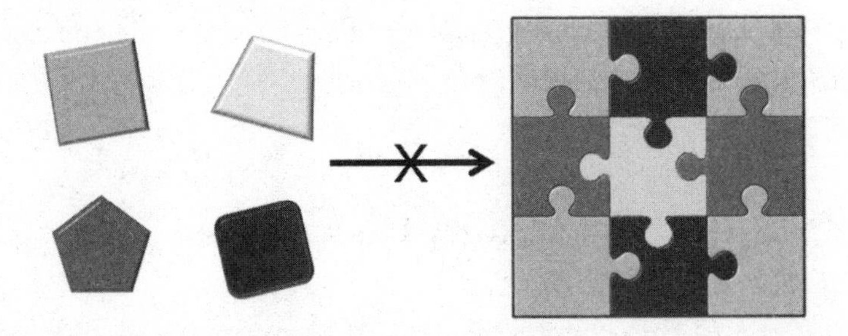

Figure 8.2. The different individual geometric shapes on the left represent individual proteins that cannot bind to one another. In order to bind, their shapes would first have to be modified into complementary forms, represented by the jigsaw puzzle on the right. This is intended to illustrate the enormous evolutionary problem of making multiprotein molecular machines, even from individual preexisting proteins.

multiple amino-acid substitutions or other mutations were needed to confer a substantially different activity on a duplicated protein, it can no longer be blithely assumed that the transition was navigated by Darwinian evolutionary processes. Some may have been, but many others not.

As I discussed in Chapter 6, new work has led me to revise the limit of organismal Darwinian evolution downward, from the biological classification level of class that I estimated over a decade ago in *The Edge of Evolution* to the level of family. On the molecular level Thornton's work forces a similar revision down from the old benchmark (two coordinated protein-protein binding sites),[15] but exactly how far down is uncertain. It may be that all proteins would encounter the same barriers as steroid receptors, or it may be that some would and others wouldn't. There's no reason to think the borderline—the edge of random evolution—that separates the planned from the unplanned has to be tidy, and good reason to think it may be jagged. Instead, as tedious as it might be, each instance, each significant new function, has to be experimentally investigated individually, in at least as rigorous a manner as for steroid receptors.

Only then can we be reasonably confident of where lie the rapidly constricting limits of Darwinian evolution.

Reality Check

For the rest of this chapter we'll turn from questions of straight biology to ones mostly about the sociology of knowledge. The main puzzle is: Why are so many Darwinists like Joseph Thornton and Richard Lenski so strangely self-assured about their theory? Over twenty years ago I devoted a chapter of *Darwin's Black Box* to surveying the evolutionary literature and demonstrating that, despite the serene confidence of many biologists, in fact there were no publications at all that described in anything like testable detail how random mutation and natural selection could account for the sophisticated molecular machinery of the cell, let alone experiments that demonstrated it. More than two decades later—despite the uproar caused by the book, despite much bluster and chest thumping in the media—the situation is unchanged. The literature remains totally devoid of explanations, and Darwinists remain incongruously smug. It seems the two have little to do with each other.

In the Appendix I revisit a few topics where my evaluation was challenged to show that it remains correct, but I won't do a similar broad survey in this book. Instead, to manifest the continuing absence of Darwinian explanations, it's enough to look more closely at the work that Thornton cites as showing its strength. He and coworkers began their 2006 paper that showed the weakening of steroid binding to receptors with a resounding affirmation: "The ability of mutation, selection, and drift to generate elaborate, well-adapted phenotypes has been demonstrated theoretically (1, 2), by computer simulation (3, 4), in the laboratory (5, 6), and in the field (7)."[16]

Clearly those citations were meant to salute the strongest relevant results of Darwinian theory before describing the group's own contribution. Of course, almost no one in the intended audience of professional researchers would be expected to actually look up the references and critically evaluate whether they did what was claimed for them. The confident assertion would be considered just an exercise in throat clearing before the paper turned to the new stuff. Yet surely newcomers to the field—eager to see how the best professionals explain exactly how Darwinian mechanisms account for the wonders of biology—would be well advised to start with them. After all, if better, more definitive work had been done, the authors could easily have cited that.

Newcomers would be very disappointed. The citations are to, respectively:

1. A book from 1930, well before the role of DNA was understood[17]

2. A theoretical study by Michael Lynch showing that neutral processes (discussed in Chapter 4) might result in one gene with two preexisting functions splitting into two genes with one each of the functions[18]

3. A computer simulation of the evolution of eye shape that ignores the role of genes, proteins, or any molecular factors[19]

4. A computer simulation of computer program development that ignores biology entirely[20]

5. A review by Richard Lenski of laboratory evolution experiments with microbes (discussed in Chapter 7)[21]

6. A study showing that one protein variant is better for bacteria under one set of growth conditions and another variant better under a second set, but neglecting to ask how either of them might have been produced[22]

7. A review by Peter and Rosemary Grant of their work with the Galápagos finches (discussed in Chapter 6)[23]

None of the works *even try* to show that random mutation and natural selection can build the complex functional molecular systems that undergird life. None even try to explain how any of the systems described in Chapter 2 or other real biological machinery could come about by undirected processes. The epidemic of tongue-tied Darwinists unable to explain how their theory might account for the real functional intricacies of life continues to this day. Better than any literature survey, that list of its purported triumphs demonstrates the impoverished state of Darwinian theory. Any claim that there is scientific warrant to believe mutation and selection can account for the foundation of life is the barest pretense of knowledge.

What Math Can't Do

So how is it that so many smart and some brilliant scientists believe, in the teeth of a barren literature, that Darwinism's most audacious, most counterintuitive claim—of being able to account for life's sophisticated structures—is well supported? Leaving aside for a later chapter the real and notoriously vexed philosophical questions, which can prod people (definitely including scientists) to take sides for nonscientific reasons, I think there are two other major overlapping reasons: (1) a socially inherited dependence on classical yet irrelevant math; and (2) the related incapacity to recognize the hardest problem of the discipline. We'll consider the first reason in this section, and the second following the next section.

The overriding role of irrelevant math is nicely illustrated by one of Richard Dawkins's objections in his 2007 review of *The Edge of Evolution* to my efforts to find the limits of Darwinian theory. I had done some elementary calculations to show that if even one step in an evolutionary pathway were not positively selected, then the wind rapidly went out of Darwin's sails, and that the problem grows exponentially worse for multiple unselected steps. Dawkins jeered: "If correct, Behe's calculations would at a stroke confound generations of mathematical geneticists, who have repeatedly shown that evolutionary rates are not limited by mutation. Single-handedly, Behe is taking on Ronald Fisher, Sewall Wright, J. B. S. Haldane, Theodosius Dobzhansky, Richard Lewontin, John Maynard Smith, and hundreds of their talented coworkers and intellectual descendants."[24] He went on to extol the great variety of dog breeds, whose degradative evolution we discussed in Chapter 7.

The first name in Dawkins's list of worthies, Ronald Fisher, whom Dawkins has called "the greatest biologist since Darwin,"[25] was also the first person whose work was cited by Joseph Thornton above. Fisher was a mathematician who turned his attention to evolution. His 1930 book, *The Genetical Theory of Natural Selection*, is considered a classic and the first major work of the neo-Darwinian synthesis. In some critical respects, though, it's now quite dated. For example, Fisher spent his first chapter, "The Nature of Inheritance," arguing why the predictions of the "particulate theory" of inheritance that follows from the work of the monk Gregor Mendel were superior to those of the nineteenth-century "blending theory." Yet what those particles (which had been dubbed "genes") were, neither he nor anyone else then knew.

Fisher was overawed by the thought that if there were only two varieties each of a hundred gene particles, the possible combinations on which natural selection could act were astronomical: "It is perhaps worthwhile at this point to consider the immense diversity of the genetic variability available in a species which segregates even for only 100 different factors. The total number of true-breeding genotypes into which these can be combined is 2^{100}, which would require 31 figures in the decimal notation."[26]

This is what Dawkins had in mind when he wrote that the rate of evolution is not limited by mutation. Because the number of combinations increases exponentially with the number of genes, if there are just two varieties of each gene, then there are four possible combinations of two genes, eight of three genes, and a billion billion trillion of a hundred genes. Since bacteria can have thousands of genes and larger animals more than ten thousand, the potential variety is immense. Given all the potential combinations to draw from, it seemed to Fisher that natural selection could pull life in any and all directions it favored.

But times have changed. Fisher didn't know what genes are. We do. Rather than amorphous, indeterminate "particles," we realize that genes are specific complex entities that code for proteins whose activities, like those of all other machines, depend on their shapes, which in turn depend on their amino-acid sequences. We understand that many mutations that change a sequence either break or damage genes, causing them to produce inactive or compromised proteins. Crucially, we also know that *broken genes can be helpful and so can be positively selected.*

This puts Fisher's calculations in quite a different light, because the two varieties of each gene he considered could simply be *a working copy* and *a damaged copy.* Ronald Fisher was right—as far as he could see. The number of possible combinations of two kinds each of a hundred gene particles is the same as he figured. What's more, that could conceivably lead to tremendous variety in life on earth, from Galápagos finches to Hawaiian fruit flies to dog breeds and more, as he thought. But he didn't grasp that the change and adaptation could be due primarily, or even overwhelmingly, to *de*volution (that is, the loss of preexisting, molecular, functional coded elements, as discussed in Chapter 7) rather than to *e*volution (the gain of such elements).

Fisher was right, but Dawkins is wrong. My calculations didn't contradict the work of either Fisher or anyone else in Dawkins's litany of the saints of mathematical genetics, because none of them even tried to explain the specific molecular machines I was discussing. Verily, *mathematics can't do that*. By itself math simply can't account for the specific physical properties of real substances. The topic lies utterly outside its domain. Math is great for rigorously demonstrating that, say, two working lawnmowers can potentially cut grass in half the time as a combination of one working and one broken mower. But pure math can never show how a lawnmower was invented.

Math, as powerful as it is when used appropriately, exists solely in the mental realm and needs to be grounded in experiments to be of any use to science in accounting for the real world. Rather than math, it is primarily *biochemistry*—the study of the concrete molecules of life—that decides what Darwinian processes can or cannot do. Only the study of the physical structures of the actual machinery of life, in all the gruesome detail necessary for their elaborate functions, can indicate whether they were produced by random mutation and natural selection.

Dawkins's servile dependence on the authority of classical math to justify Darwin is a stunningly naive category error. It's as if he had tried to explain radioactivity by discussing iambic pentameter. Yet for forty years he has been Darwinism's most celebrated popularizer—he was even elected by Britain's top scientists as a fellow of the prestigious Royal Society. Since he has been so widely feted by them, it's reasonable to think his elementary confusion is widely shared by evolutionary biologists. That's appalling—but not surprising. The original misunderstanding likely started benignly enough in Ronald Fisher's day, when genes were unknown abstract entities, but then was passed down over the generations to the present, from professor to graduate student, as a pernicious, unchallenged, largely unrecognized assumption, even as the biological ground shifted radically, until the whole field has reached the pitiable state where Joseph Thornton's list of references passes as amazing evidence for the power of mutation and selection.

Damaged Genes Can Drive Diversity

Even though it doesn't clarify the underlying basis for itself, the mere fact that species such as Galápagos finches and African cichlids can diversify in different environments understandably impresses many people, who then mistakenly attribute vast constructive powers—rather than merely adaptive ones—to Darwinian processes. To try to distinguish more clearly between the two, here's a folksy analogy to show how the *loss* of an ability can drive specialization.

Suppose in the early 1800s four pioneers separately departed Philadelphia to settle in distant regions of America. Among other supplies, from the local general store each purchased the same tools: a canteen, compass, rifle, and sleeping bag (Fig. 8.3). Because of poor manufacturing quality control in those days, however, for each pioneer a different one of the four tools was defective.

How might that affect the pioneers' destinations? Well, the one who had a broken canteen might decide to settle near a river, so he'd always have a ready supply of water. The one missing a sleeping bag might settle where it was warm, to lessen nighttime exposure. Since he couldn't navigate easily, the one with a broken compass might stay close to where he started. Since he couldn't hunt, the one with a defective rifle might decide to look for good farming land.

Because each was missing a separate tool, they diversified in different environments where they could more easily thrive. It was the very *lack* of tools—not the discovery of new ones—that matched them with their surroundings. Yet, of course, all of the tools were already available in the general store—the pioneers invented none of them. The fact that the absence of a tool drove them to different environments says nothing at all about how the tools originated.

Switching back to biology, consider two varieties of the gene for one of the chains of hemoglobin. One variety is the kind most frequently found in humans. The other has lost a quarter of its length

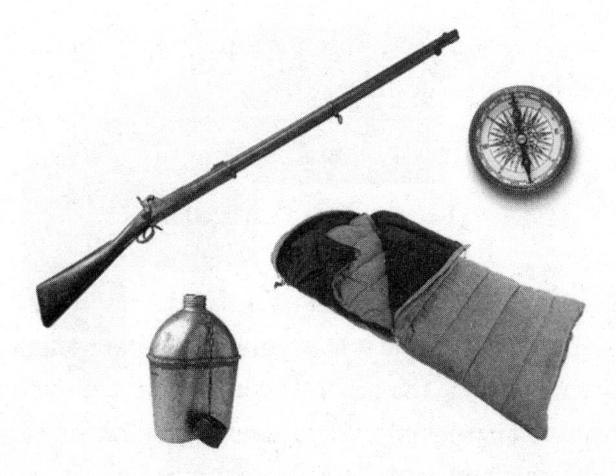

Figure 8.3. Pioneers missing different tools might settle in alternative environments, as discussed in the text.

through deletion, rendering it nonfunctional, as happens in some people with the blood disorder thalassemia. And consider two varieties of the gene APOC3, one of which is the normal functioning gene and the other is a broken copy, which, as we saw in the last chapter, confers resistance to heart attacks on some people. Thalassemic people survive more successfully in malarious regions of the world. People with broken APOC3 genes can eat the high-fat diet of modern society with less risk. Maybe people with a combination of the two mutant genes could eat a high-fat diet and thrive in a malaria-infested region.

Does the existence of the two forms give us any idea of how random mutation and natural selection could produce hemoglobin in the first place? Of course not, no more than comparing a working auto with a junkyard wreck tells us how cars were invented. Does the existence of the two forms of APOC3 give us any idea how random mutation and natural selection could produce the gene in the first place? Again, no. It just shows how broken genes can be fodder for natural selection. In other words, mutation and selection can indeed produce a wide variety of beneficial traits that fit organisms to different environments—perhaps as many as Ronald Fisher thought—and yet still show absolutely nothing about how the underlying machinery of life arose.

Genes as Widgets

Now let's briefly consider the second, related, reason why so many biologists mistakenly consider Darwinism to be well supported: the incapacity to recognize the hardest problem of the discipline.

Like evolutionary genetics, economic theory can employ a lot of sophisticated mathematics. Because they are professionally interested in the trading of goods, not in the goods themselves, economists often write about generic imaginary products they call widgets. The word "widget" acts as a placeholder in their theorizing. So if a factory has the capacity to make a thousand widgets per month, and demand depends on the unit price of the widget according to some formula, theory can calculate the optimum number of widgets up to a thousand per month that the factory should make to maximize its profit. Given the cost of building a new factory and perhaps a few other pieces of information, one can even calculate when it would be optimal to build a new factory if demand exceeds capacity. The calculations can work well in reality whether "widget" stands for a cell phone, a tractor, or a vacuum cleaner.

For theoretical evolutionary biology, genes are widgets—that is, mainly featureless abstract entities whose behavior can be described mathematically. For example, Ronald Fisher calculated the number of combinations of a hundred genes with little knowledge of what they were. Beginning students of evolution are quickly taught about the Hardy-Weinberg equation, which states that in the absence of evolutionary influences the frequency of two different forms of a gene, called *alleles* (like the two varieties that Fisher considered), in a population will remain the same over generations, no other details necessary. If the effects of an allele appear to be helpful or detrimental, then a factor called the selection coefficient can be added to the equations to accommodate it. If the population size of the species carrying the gene increases or decreases, if two different alleles seem superior to two copies of the same, if the species reproduces clonally or sexually or a combination of the two, then the equations can be adjusted to model all those effects. Yet it matters little to the theory which physical structures the genes encode.

The theoretical/mathematical emphasis leads inexorably to a view of genes as just so much putty in the hands of abstract forces that are the real explanation for life. With little more than a gesture at astonishingly sophisticated molecular machinery, equations on a sheet of paper are taken by many as a sufficient account. For the many biologists such as Richard Dawkins who can't themselves do the math, the very existence of the classic texts reassures them of the unshakeable validity of neo-Darwinian theory. This leads to a vicious cycle in which theoreticians calculate without regard to the particularities of genes or proteins, experimental biologists trust that whatever particulars they do discover can be accounted for by the theorists, and each group reassures the other that all is well. Even when lab researchers such as Joseph Thornton and colleagues encounter surprising, objectively problematic results lurking in the complex structures of proteins, they're written off as anomalies. Ronald Fisher's nearly infinite number of possible variations is imagined to rescue the situation. Thus the hardest problem of biology—how to explain the origin of the particular sophisticated functional structures of life—is effectively rendered invisible.

Of course, sharp criticism from outside the charmed circle seems bizarre at best, since virtually everyone in the field agrees on the basics, so all radically skeptical arguments by interlopers are chalked up to ignorance or bad faith. Yet so ingrained is the groupthink that even modest criticism from within the charmed circle seems incomprehensible. Masatoshi Nei (mentioned in Chapter 4) wrote his 2013 book *Mutation-Driven Evolution* in order to highlight the breathtakingly obvious point that particular defined mutations are needed to account for the concrete molecular features of the cell; hand-waving stories of selection acting on generic mutations simply beg the question. The book received respectful but puzzled reviews: "What is remarkable is that the author is not someone from the fringes or even outside of evolutionary biology. . . . [He] is one of the founding fathers and pioneers of what is now called the field of molecular evolution. . . . Isn't selection well studied and the well-established driver of adaptive evolutionary change?"[27] Another reviewer had to reassure his readers, "Nei does not deny the existence of natural selection."[28]

The study of economics is useful, but it's ludicrous to think the law of supply and demand called forth the existence of, say, microwave ovens. If academic economists in public, in the light of day, discounted the inventiveness of the engineers whose efforts actually yielded the marketable products, to that extent we would rightly judge them to be disconnected from reality. No law explains the commercial products whose trading is studied by economics.

Unfortunately, to a very large extent evolutionary biologists do think the most marvelous molecular machinery is called forth by natural selection acting on theoretical shapeless, generic mutations. Regrettably, for all practical purposes, mathematical geneticists do discount the inventiveness displayed in life, in the sense that they don't take it seriously. To that extent, the whole field of evolutionary theory really is disconnected from reality. No law explains the molecular machinery whose descent is studied by evolutionary biology.

One to Go

In this and the previous chapters we considered two factors that help make evolution self-limiting: random mutation and natural selection—in other words, Darwin's mechanism itself. In the next chapter we look at the final one—irreducible complexity—and at how the three factors interact with each other.

CHAPTER 9

Revenge of the Principle of Comparative Difficulty

It's taken a century and a half to realize it, but random mutation and natural selection are self-limiting. Two-edged swords, they both promote Darwinian evolution on a small scale and hinder it on a large one. As we've seen in the previous several chapters, selection fits a system more and more closely to its current biological task, just as we expected, but that makes it more and more difficult to adjust to other potential functions, which we didn't. Random mutation supplies beneficial variation, as we were taught, but it comes predominantly at the expense of a species's store of genetic information, which we weren't. The effectiveness of Darwin's mechanism on a limited scale can be seen with just a sharply observant naked eye, available in the nineteenth century. It's foundering at greater scales could only have been discovered after the biological and computer revolutions of the past sixty years.

In contrast, the problem of functional complexity was clear from the beginning. Actually, even before the beginning. More than fifty years before Darwin's theory, William Paley cited the eye as a dazzling example of purposeful design, likening it to an intricate watch: the clever arrangement of their components required intelligent direction as much in the one case as in the other. The most potent early attack on Darwin's theory after its publication was *On the Genesis of Species* in 1871 by St. George Mivart. A keystone of his argument was that the beginning stages of complex structures such as the eye would have had no use and therefore would not have been selected. Sure, Mivart and other skeptics agreed, once a structure was in place, Darwin's mechanism could kick in and modify it to a greater or lesser extent. But formation of the initial system was beyond its capabilities.

Darwin shrugged off the problem. After all, speculating then about how vision arises was futile: "How a nerve comes to be sensitive to light hardly concerns us more than how life itself first originated."[1] In the *Origin of Species* he gestured at some modern animals with simple eyes and others with more complex ones and yawned that he had no worries: "I can see no very great difficulty . . . in believing that natural selection has converted the simple apparatus of an optic nerve . . . into an optical instrument as perfect as is possessed by any member of the great Articulate class."[2] Darwin's languid response didn't rise even to the level of a "Just So" story, yet his modern defenders treat it as the next best thing to a demonstration. Despite the profound progress of molecular biology in the meantime, the 150-year-old caricature of a simple flat eye curling up to give a round vertebrate-like eye has been repeatedly invoked by his most prominent advocates.[3]

Despite their compelling prima facie challenge to Darwin's theory, discussion of the evolution of complex anatomical systems such as the eye can easily lead to confusion, because what Paley and others treated as single components—the retina, the lens, and so on—are actually themselves stupendously complex aggregate systems composed of many kinds of active cells and molecules. In Paley's and Mivart's days, the existence, let alone the abilities, of the sophisticated molecules that fill cells was unknown. They were all unwittingly lumped under the term "protoplasm"—seemingly just a gooey nondescript jelly. Without a good understanding of its nature, who can tell what mysterious protoplasm might do? For all the best minds of the mid-nineteenth century knew, it might stretch and shape itself into anything. When ignorance reigns, it's as easy for a dreamer to imagine a simple eye morphing into a complex one as to imagine that "gemmules" explain heredity.

Mutations—the raw material of evolution—are changes in *molecules* (in DNA and proteins). Cells—the fundamental level of life—are built of molecular machines and molecular systems. So, to avoid confusion, discussion of evolution has to focus on life's foundation—the *molecular* level. When we do, it turns out that we unambiguously see the same conceptual difficulty for Darwin's theory there that William Paley and St. George Mivart saw at the organismal level of life; that is, the cell too is chock-full of elegant machinery that requires multiple interacting components. But now there's no mysterious lower level of life for imagination to retreat to.

In the first part of this chapter I briefly describe the problem of irreducible complexity, which I treated at length in *Darwin's Black Box*, and then show that the difficulties it presents for Darwin's theory have grown much worse in the past several decades. We'll then explore how the three factors of random mutation, natural selection, and irreducible complexity reinforce each other to ensure that Darwinian evolution is self-limiting.

Irreducible Complexity

In the *Origin* Darwin insisted that evolution as he envisioned it had to occur slowly, in tiny steps, over long periods of time. He realized that if helpful coherent complex changes arose quickly, in large leaps, then something other than random variation would have to be involved. "Natural selection can act only by taking advantage of slight, successive variations; she can never take a leap, but must advance by the shortest and slowest steps."[4] Thus complex structures posed a potentially ruinous problem: "If it could be demonstrated that any complex organ existed, which could not possibly have been formed by numerous, successive, slight modifications, my theory would absolutely break down. But I can find out no such case."[5]

To help ensure that no such case ever was found out, notice that Darwin cleverly foisted on critics a hopeless standard to meet—to "demonstrate" that something "could not possibly" happen. In other words, to prove a negative. Science, of course, cannot prove anything to be logically impossible, most especially if imaginative stories count as support. The rhetorical ploy was understandable at the time, since Darwin wanted to shield his infant theory from being dismissed out of hand, without a proper hearing. Nonetheless, the problem of complex structures is a very real one, and his theory is all grown up now. So let's ignore Darwin's unrealistic, defensive standard and just ask what sort of complex organ or system certainly doesn't look like it could be put together by random changes and natural selection in "numerous, successive, slight modifications."

A kind of system that strongly challenges Darwin's mechanism is one that is *irreducibly complex* (IC). In *Darwin's Black Box* I offered a working definition: "By *irreducibly complex* I mean a single system composed of several well-matched, interacting parts that contribute to the basic function, wherein the removal of any one of the parts causes the system to effectively cease functioning."[6] As an illustration of irreducible complexity from our everyday world, in 1996 I pointed to a common mechanical mousetrap of the kind one can buy in many grocery stores (Fig. 9.1). A mousetrap consists of a number of pieces. It has a large wooden base to which everything else is attached. There is a tightly coiled spring with extended ends that press against the base and also against another

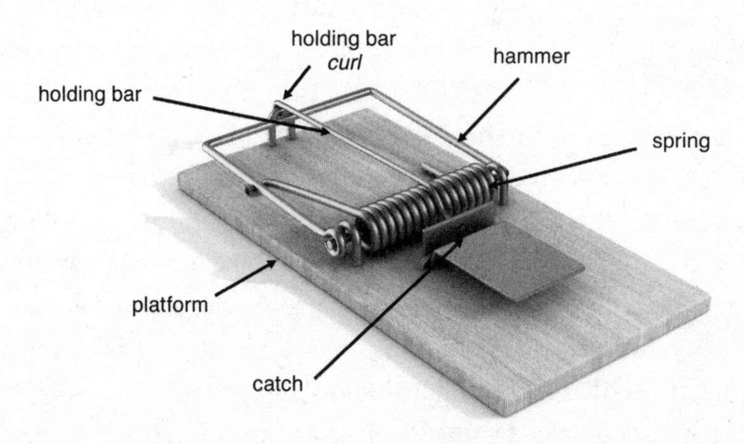

Figure 9.1. A common mechanical mousetrap needs multiple pieces that are themselves complex.

metal piece called the hammer. The hammer has to be stabilized by a piece called the holding bar to keep it in position. And the far end of the holding bar itself has to be inserted into a piece called the catch. Besides these major pieces, there are assorted staples that attach them to the base.

How could something like a mousetrap evolve gradually by something like a Darwinian mechanism, by "numerous, successive, slight [and, Darwin neglected to add here, *random*] modifications"? A wooden base alone wouldn't catch mice, so natural selection would have nothing to select at that point. Even a base with, say, several of the staples in place or with the holding bar attached still wouldn't function as a trap. The general barrier IC presents to Darwin's gradual mechanism is that if a system requires a number of components for its function, then natural selection cannot favor the function until all the needed pieces have already come together. In other words, the system first has to exist before selection can affect it. That's just another way of stating St. George Mivart's problem of the incipient stages of complex structures. The predicament is easy to see.

Over the years a veritable cottage industry of Darwin defenders has strained to discredit the mousetrap in scholarly books,[7] journals,[8] and internet postings.[9] Despite such efforts it remains the reigning paradigm of irreducible complexity. In 2004 I rebutted at length what I considered the most interesting objections to IC.[10] I won't rehearse all those arguments here. Rather, I'll just mention the most common one: many critics reprise Darwin's own gambit, to force skeptics like myself into trying to prove a negative. "Demonstrate," they insist, that a functioning trap "could not possibly" arise gradually; prove that it's somehow logically impossible.[11]

But that's a completely unsuitable standard. Although it uses logic, science judges the success of a theory by the weight of empirical evidence. The appropriate straightforward criterion is this: if there are good physical reasons to think Darwinian routes wouldn't work and if after a diligent search no evidence is found that they do, then the theory has failed. There's no obligation to pretend otherwise, no requirement to hunt forever for the Loch Ness monster.

Here's a critical implication of the mousetrap problem: if even comparatively simple machinery is irreducibly complex, then, except for the very simplest (such as, say, an inclined plane or basic wheel), pretty much all other machinery is too (Fig. 9.2). Of course complex automobiles and air conditioners and sewing machines need multiple well-matched parts to work. But so do less complex bicycles and push lawnmowers and tire jacks. Yet one of the major discoveries—arguably *the* major discovery—of modern biology is that the cell is run not by amorphous protoplasm, but by discrete complex *machines*—literally, machines made of molecules.[12] There are molecular machines such as the famous bacterial flagellum that act as outboard motors, others that work as trucks to ship supplies throughout the cell, still others that act as traffic lights, road signs, and more.

All even moderately elaborate machinery is irreducibly complex. The machinery of the cell—such as described in Chapter 2, in my

Figure 9.2. A complex gearbox. If a simple mousetrap is irreducible, so is virtually all complex machinery.

previous books, or in any basic biochemistry textbook—is very elaborate indeed. Therefore it too is irreducibly complex. Since IC systems are quite resistant to gradual construction by an unguided process such as Darwin's mechanism, and since there is no plausible evidence to show that they can be so constructed, it is reasonable to conclude from that alone that random mutation and natural selection did not produce the molecular machines of the cell. What's more, as we'll see below, the actual situation is much worse. When we leave imaginative scenarios behind, in the real world Darwin's mechanism has profound problems even with biological features that are much simpler than a mousetrap.

For decades opponents of intelligent design have tried and failed to find a plausible Darwinian route even to a simple mechanical mousetrap. As I show for some representative cases in the Appendix, for the most part they haven't even tried to do so for the very complex molecular machinery of the cell. Although it can nicely explain *some* biological adaptations, such as sickle-cell hemoglobin or the varieties of cichlid fish, Darwin's theory has failed for molecular machinery.

Complex Parts

A mousetrap doesn't consist just of a handful of nondescript parts. The parts themselves have multiple features that must be the proper size, shape, and strength and be positioned correctly for the system to work. The extended ends of the spring have to be oriented in the right directions. The staples have to be placed at the correct positions. The spring has to have the right rigidity. Notice in Figure 9.1 that the holding bar has a little curl at one end to grasp a staple that attaches it to the base. Without that curl, the whole trap fails. A realistic discussion of the difficulties for Darwinian-like evolution would have to consider the myriad details of the trap components, not just the number of them.

As helpful as it is as an illustration of irreducible complexity, an ordinary mousetrap can obscure the immensity of the task facing the undirected evolution of life. The reason is that, more so than mousetrap parts, which have multiple necessary details, the protein parts of molecular machines are themselves extraordinarily complex. For example, we might speak of hemoglobin as a single protein that binds oxygen and delivers it to cells. But that one "part" consists of over *five hundred* amino-acid residues whose sequence has to be very sharply specified in order for it to work (Fig. 9.3). Recall that Joseph Thornton's studies identified a couple of problematic amino-acid residues that blocked steroid-receptor evolution. Yet that class of proteins consists of five hundred to a thousand residues. Roles that are simple to describe—"binds oxygen," "binds steroids"—actually require elaborate structures to carry them out. In this and the next few sections we'll briefly look at some of the excruciating level of detail that is necessary for life to work.

Let's call complex systems with parts that are themselves complex *comprehensively* complex. All molecular machinery is comprehensively complex, even if the role of a particular part in a system

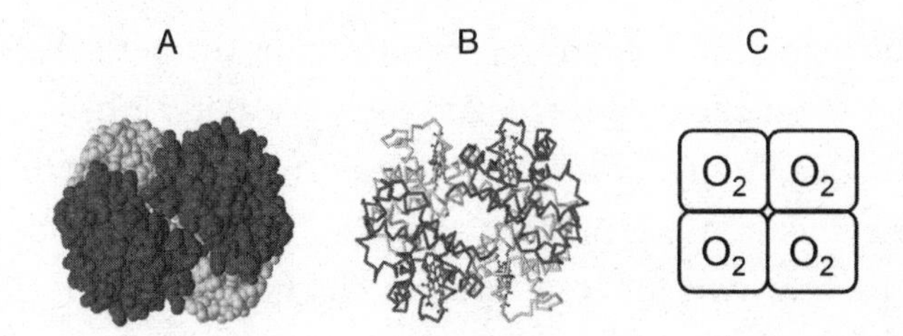

Figure 9.3. Hemoglobin simplified. To highlight various features, different renderings of a protein can show different amounts of detail. Yet life requires *all* of the detail. (*A*) A space-filling model of the thousands of atoms of hemoglobin. (*B*) A less detailed model with line segments connecting Cα-carbons of successive amino-acid residues. (*C*) A simple cartoon depicting the four subunits of hemoglobin as geometric squares, each of which can bind one oxygen molecule.

is *conceptually* simple. For example, suppose that for some purpose a cell needed a protein part simply to assume a rigid L-shape, with one side twice as long as the other. That would require an initially floppy chain of amino acids to fold upon itself so that electrostatic forces between residues would interact to supply the needed rigidity in the correct proportions to yield the right ratio of lengths. In turn, that means the sequence of amino-acid residues in the protein would have to be correctly ordered, so that the right positions in the chain attracted the right complementary positions that would lead to the L-shape. Without the correct order of perhaps hundreds of amino-acid residues, the "simple" part would fail.

Even conceptually simple protein parts are comprehensively complex, but of course most molecular machines are not even conceptually simple. One of my favorites is called gyrase, an enzyme that has the amusing ability to tie DNA into literal knots. (Its practical use to the cell is to "supercoil" DNA—twisting it up like an overwound rubber band and in the process storing mechanical energy in the very shape of the DNA itself, which can later be used for various purposes.) To do so the machine has to grab two separate places on each strand of double-stranded DNA, chemically break one of the two on each strand, push the intact region through the break by coupling to an energy source, and then reseal it. Unlike our simple L-shape illustration, gyrase has to perform all manner of dynamic rearrangements. Yet it manages to do all of its tricks with just two copies each of two different protein chains—the same number as hemoglobin. It can do that because its active features are all within the chains. Gyrase is an example of an intricate molecular machine whose severe challenge to Darwinian evolution is hard to frame in terms of "parts" and irreducible complexity. Its approximately *three thousand* amino-acid residues, however, are a terrific example of comprehensive complexity.

Mini–Irreducible Complexity

Now let's look even closer at some necessary details of proteins. An immediate implication of comprehensively complex molecular machines such as gyrase is that even single proteins are collections of many necessary features. Like everything else about a protein, those features depend on multiple particular amino-acid residues in particular positions interacting with each other in particular ways. Perhaps the simplest example of a feature needing multiple amino-acid residues is called a *disulfide bond* (sometimes spelled *disulphide*), which is a chemical link between two amino-acid residues called cysteines that acts kind of like a hook-and-eye latch (Fig. 9.4). Neither a hook nor an eye by itself can fasten a door, and neither can one cysteine by itself form a disulfide bond—it requires two. A more complex example is a binding site, either for another molecule on the surface of a protein or simply for another section of the same protein in its interior. A binding site is necessarily composed of multiple amino-acid residues that match the shape and chemical properties of the region it is to bind.

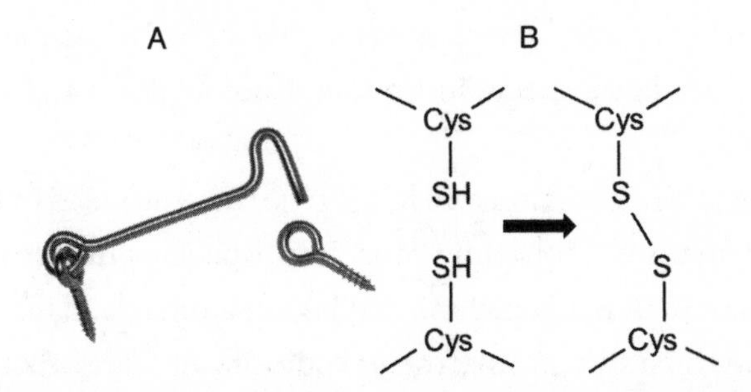

Figure 9.4. Even the simplest mini–irreducibly complex features are huge headaches for Darwinism. (*A*) A hook-and-eye latch. (*B*) Two cysteine groups forming a disulfide bond.

On a much finer scale than the examples I discussed in *Darwin's Black Box*, structural and functional protein features that require multiple amino-acid residues call to mind irreducible complexity. That is, they require multiple parts to work (the amino-acid residues of which they consist) in addition to the general structure of the protein in which they are embedded. If the parts aren't there, the feature simply does not exist. That makes the features effectively impossible to evolve in a gradual fashion. Imagine, for example, that it would be beneficial for a protein that didn't already have one to evolve a disulfide bond. The first cysteine that appeared by random mutation wouldn't make the feature—only when the second one appeared could the link form.

Since they are akin to irreducibly complex systems but on a smaller scale, let's dub them *mini–irreducibly complex* (mIC) features. A difference between mIC and full-blown IC systems is that mIC features don't have stand-alone functions, but are aspects of larger systems that do. Examples from our larger world are a hook-and-eye latch (which can keep a door shut), the teeth of gears (which mesh with other gears in, say, a watch), the cuts in a key (which help open a lock), and so on. Because mIC features need multiple parts, we might suspect that their evolution too will present a hurdle to

Darwinian processes—perhaps not an individually insuperable one, but still a very large speed bump that would substantially interfere with it.

In the next three sections we have to go into the weeds a bit to consider some math and computer modeling aspects of mIC features. But the excursion will really pay off. We'll gain a reliable, definitive yardstick with which to measure the grandiosity of Darwinian claims.

Modeling mIC Features

The notion that several amino-acid mutations would have to appear before a particular selectable feature formed in a protein is really a pretty elementary idea, an obvious inference from the structures of proteins that have been known since the mid-1950s.[13] Nonetheless, no one seems to have systematically investigated the problem before University of Pittsburgh physicist David Snoke and I published a study of it in 2004.[14] Our work developed a computer model that calculated how many generations on average a species would have to wait before the multiple mutations to form an mIC protein feature requiring two, three, four, or more changed amino-acid residues would be expected to arrive.

In models of protein evolution a few factors at least can be relied upon, specifically the general DNA mutation rate and the general likelihood that a mutation will damage a protein's ability to function. The exact sequence of postulated events, however, is speculative. Whether computer mutations are considered to be helpful, harmful, or neutral; whether mutations are limited to point mutations, gene duplications, deletions, or insertions; whether the model takes into account recombination of genes in sexual reproduction—all of these and more are arbitrary decisions of the investigators. By far the most important feature of a computer model for how the modeled system is expected to behave is the starting assumptions. For example, if a long-term model of the economy assumes that, say, the relative price of health care will stay constant, then the model may be very misleading if medical costs fluctuate. The more uncertain the starting assumptions and the longer the term it tries to account for, the less reliable the model. Those caveats should be kept in mind for all computer studies of evolution.

Snoke and I set out to test how quickly mini–irreducibly complex (in the paper we called them "multiresidue") features could pop up in a simple conceptual model of protein evolution that had been used earlier by prominent mathematical geneticists.[15] Very briefly, a hurdle to the development of a new function by an old protein is that a mutation could well destroy the original activity of the protein, which might still be needed by the cell. So, like those earlier workers, to circumvent the difficulty we assumed the gene for a needed protein had accidentally duplicated. In that case one of the two copies of the duplicated gene could continue to fulfill the original function, while the other copy could randomly accumulate mutations, perhaps leading to a new function.

Because so many positions are needed for it to work, the great majority of mutations to a duplicated gene would be expected to damage the protein it encoded, preventing it from developing a new feature. The more mutations that had to accumulate for the new feature, the much less likely was success. Nonetheless, the more generations that passed and the greater the number of organisms in the species, the greater the chances that the constructive mutations would eventually arrive before a damaging one did. The point of the model was to use reasonable assumptions to estimate how long that process would likely take. The bottom line was that it would take a very long time indeed to develop even the simplest, two-amino-acid mIC feature. The situation for more complex features grew worse exponentially.

When we completed our study, David Snoke and I thought it would be of some interest to other biologists, so we wrote up the results and submitted the manuscript to a journal called *Protein Science*. The goal of the fourteen-page paper was not to "disprove" Darwinian theory, as some excitable folks then thought (see next section). Rather, the modest goal was to show that one common model for protein development—gene duplication followed by accumulation of mutations to yield a new function—was implausible if multiple changes were needed.

The manuscript was reviewed according to the journal's standard protocol, revised to accommodate the reviewers' comments, and published in the October 2004 issue with the soporific title "Simulating Evolution by Gene Duplication of Protein Features That Require Multiple Amino Acid Residues." It attracted some attention.

Premises, Premises

Emails soon flooded the journal editorial office demanding an explanation for the journal's publication of a paper by a known intelligent-design advocate (*c'est moi!*). Journals rarely receive comments on articles they publish, especially not angry ones, so the staff was nonplussed. I won't go into the commotion. The important result is that within a bare week of the appearance of our paper the journal editor notified us that mathematical geneticist Michael Lynch (whose work on neutral theory I discussed in Chapter 4) had submitted a manuscript to rebut our work. His paper was subsequently reviewed, revised, and published the next September.[16] We were permitted a thousand-word reply.[17] Despite the swirl of events, the outcome is pretty close to the ideal of science. A study is reported on an important issue, and those who are doubtful of the results take their best shot at calling them into question. If the work can't stand up to the heat, better that should be known quickly. If the strongest criticisms of knowledgeable and dedicated opponents don't topple it, however, its credibility is strengthened.

Lynch offered a model with different premises. Instead of assuming, as we did, that most amino-acid mutations damage a protein's function, he assumed needed changes were neutral—neither helping nor hurting. The assumption of neutrality allowed the first mutation to appear in a working protein, which immunized it to the inactivating mutations that were possible in the model we proposed. Lynch also postulated that any new mIC feature would replace the original function of the protein, so its gene had to duplicate before the final step to allow one copy to retain the old function and the new copy to take up a novel one. (Our model assumed the gene duplication came first.) He restricted his model to considering only the minimal number—two—of amino-acid changes; we had investigated multiple changes. He also allowed up to fifty different possible locations in a protein chain to give rise to the changes; we had specified that each could occur in only one position (because the exact position of a feature within a protein is often critical to its function). In short, Lynch's model had a number of features that differed from ours, but still arrived at the same destination—a duplicated protein with a new multiresidue feature.

And despite Lynch's emphasis on the shorter times his model predicted compared to ours, both models showed the same qualitative pivotal, bottom-line result: if just two mutations are needed to get some selectable effect, that requires either much longer times, much greater population sizes, or a combination of the two. In other words, the difficulty of producing even the *simplest* mIC feature is enormously greater than for one mutation. As an example, for a population size of a million organisms, if gene duplication isn't needed, it's expected to take about ten thousand generations to mutate just one particular amino acid in a particular protein.[18] To produce a feature in a duplicated gene that needs two such mutations, in Lynch's model a *hundred million* generations are needed.[19] Our model indicates about a *billion*.[20] A hundred million is of course much less than a billion, but both numbers are much, much greater than that needed for a single mutation.

In a computer one can always manipulate the expected time to a mutation by assuming the hypothetical population size of a theoretical species to be larger or smaller, the target region of a gene to be greater or smaller, or the helpfulness of the new feature to be stronger or weaker. Lynch's paper emphasized optimistic cases of all those variables. But none of the factors alter the bottom line that two required changes are enormously more difficult to obtain by random mutation than one. And when a very intelligent critic dedicated to proving something wrong comes up with at least the same qualitative behavior, you can bank on it being correct.

If just two simple molecular changes are needed for a feature to evolve, there's a quantum leap in difficulty for Darwin's mechanism. The more required changes, the exponentially worse it becomes. As I'll explain in the section after the next, that's an insurmountable problem for undirected evolution, but not primarily because of the amount of time involved. Rather, it's fatal because *damaging* a gene only requires a *single* hit, and it is the *ratio* of times that is crucial. Since single mutations will appear so much faster, that means the kind of damaging yet beneficial mutations revealed by modern research will spread in a comparative lightning flash, ages before the completion of any mIC feature. Poison-pill mutations will *always* dominate a Darwinian evolutionary landscape.

"The Old Enigma"

Dave Snoke's and my paper attracted much criticism. Yet when the same topic is raised by less controversial figures, the reception can be quite different. To understand the formidable challenge posed by mIC, it's a lot more revealing to read exultant papers in which a solution is thought to have been found, rather than defensive responses to critics.

Eugene Koonin (whose work I discussed in Chapter 4) is a terrific scientist, prolific author, and booster of Michael Lynch's idea that neutral processes account for many features of eukaryotic genomes. Koonin is also the editor of a journal called *Biology Direct*, one of a number of new online journals set up to handle the burgeoning flow of results from genetics. An unusual feature of the journal is that it publishes the names and comments of the reviewers of a paper. So readers get to see any thoughts those reviewers may have had.

In 2008 the journal published a paper (entitled "The Look-Ahead Effect of Phenotypic Mutations") that concerned a fascinating and familiar problem: "The evolution of complex molecular traits such as disulphide bridges often requires multiple mutations. The intermediate steps in such evolutionary trajectories are likely to be selectively neutral or deleterious. Therefore, large populations and long times may be required to evolve such traits."[21] Notice that's *exactly* the problem Snoke and I addressed in our paper. From reading the many responses to our work, however, you'd guess that getting multiple coordinated mutations was something everyone already knew ordinary Darwinian processes could easily do. But that's not what Eugene Koonin thought. Acting as one of the reviewers himself, he wrote: "The idea of this paper is as brilliant as it is pretty obvious . . . in retrospect. A novel solution is offered to the old enigma of the evolution of complex features in proteins that require two or more mutations (emergence of a disulphide bond is a straightforward example). . . . From my perspective, this is a genuinely important work."[22]

"Old enigma"? *"Old enigma"*? Who knew that getting just two coordinated point mutations by random mutation and natural selection was a long-standing mystery?

The point I want to make here is not about the crippling deficiencies of this or that hoped-for solution, such as the authors of the paper proposed.[23] Rather, the overwhelmingly important, completely overlooked point is the degree to which Darwin's theory struggles to account for even the *simplest* example of even a *mini*-IC feature. Recall that the thoroughly obvious problem of complex interactive structures such as the eye was pointed out by the biologist St. George Mivart soon after Charles Darwin wrote the *Origin of Species*. Darwin waved it away as a problem for the future. Well, the future has arrived. The future studies (from Darwin's perspective) detailed in the past three sections show that Mivart was more right than he knew: even the *slightest* need for coordination—let alone for the immense organization of an eye or life in general—has the theory panting as heavily as if it had just tried to climb Mt. Everest. The palpable excitement and relief in Koonin's voice show much less about the paper he was describing than about the importance and intractability of the problem it attempted to address.

Uncle Rico

So, in light of its struggles in tiny matters, how should we think of the grand claims for Darwinism? Analogies can help. The claim that Darwin's theory explains life is like the claim that an illiterate who doesn't know that *u* follows *q* authored *Romeo and Juliet*. It's like a guy who says he's an Olympic hurdler, but can't lift his foot over a curb without tripping. It's like saying the theory can easily explain an outboard motor—it just has trouble explaining the hook-and-eye latch on the shed where it's stored. It's like Uncle Rico in *Napoleon Dynamite* asking, "How much you wanna make a bet I can throw a football over them mountains?" It's like . . . Well, you get the idea. No unaccomplished braggart in the world could match Darwinism's record.

Back to biology. With its boosters blissfully ignoring the Principle of Comparative Difficulty, the theory that labors mightily to explain a crummy two-amino-acid-residue disulfide bond and that has trouble trying to account for the most trivial changes in steroid receptors is ludicrously claimed to account for:

The entire five-hundred-residue steroid receptor itself

The astounding gyrase, with its three thousand residues

Coordinated multiprotein systems such as the blood-clotting cascade

Molecular machines such as the bacterial flagellum

Intricately coordinated gene regulatory networks

The differences between retinal cells and muscle cells

Organs of extreme perfection such as the eye

The ability of cells to form coherent organisms such as flies and frogs

Those bizarre claims would elicit our pity if they were shouted by someone dressed as Napoleon in the town square. That they are asserted, often belligerently, by some of the world's most intelligent scientists shows that craziness is not at all confined to the clinically insane. Rather, as G. K. Chesterton observed, craziness comes from obsessing over one idea: "Such is the madman of experience; he is commonly a reasoner, frequently a successful reasoner. . . . He is in the clean and well-lit prison of one idea."[24]

And, most regrettably, this mental disease is contagious. When the leaders of a central scholarly discipline such as biology obsess over one idea, it drags down much of intellectual life with it. Perhaps you have read that Darwin's theory also explains politics,[25] the law,[26] literature,[27] music,[28] love,[29] the universe[30]—even mind itself.[31]

It just has trouble accounting for a disulfide bond.

All the Time in the World for *De*volution

In this and the previous several chapters we've individually discussed three hurdles to long-term Darwinian evolution: random mutation, natural selection, and irreducible complexity. Now let's consider how they interact with each other. As mentioned earlier, random mutation and natural selection both promote evolution on a small scale and hinder it on a larger one. Mutation supplies the variation upon which natural selection acts, but the greatest amount of that variation comes from damaging or outright breaking previously working genes. In the case of an already functioning complex system, natural selection shapes it more and more tightly to its current role, making it less and less adaptable to other complex roles.

Notice that those two factors inhibit evolution in different ways. The degradation of genetic information caused by random mutation is a separate problem from the stultifying grip of natural selection. Because they operate through different independent mechanisms, the problems caused by mutation and selection are multiplied. As an illustration, if the odds of winning a small prize in the weekly state lottery were 1 in 100 and the odds of winning the weekly church bingo game were 1 in 200, then the odds of a particular person winning both in the same week would be the multiple of those, 1 in 20,000. The independence of the two factors makes them considerably more potent than they otherwise would be.

Irreducible complexity is an additional independent hurdle for evolution. The need for multiple coordinated mutations is a different problem from the degradation of genetic information or the constricted fit of a system to its current role, so the problems it presents also multiply the troubles for Darwinism. But irreducible complexity is much more than just another independent problem. Like placing normally far-separated hurdles for a track meet a few yards apart so that a runner jumping over one would crash into the next, mIC synergistically aggravates the problems caused by mutation and selection, making the situation for Darwin much worse—that is, much worse even than what you'd expect from just multiplying all of the troubles together. The reason is that mIC features require much more time to form than do simple single changes, so the time available for random mutation to cause—and natural selection to spread—mischief is greatly extended. Thus whatever selective pressures a species experiences will be alleviated by quick, damaging fixes well before any otherwise helpful, constructive mIC feature arrives on the scene.

Here's a fanciful illustration. Suppose you lived in a crude walled area on a hillside. Persistent heavy rains have recently led to water accumulating inside the walls and rising at the rate of a foot per day. You, who are under 6 feet tall, have less than a week to solve the problem before you drown. One possible solution is to build a mechanical pump to eject the water. How to do so? Once a day on average random debris from outside is blown over the wall into your compound; perhaps you could wait until there are enough pieces of debris that could be fit together into a pump. The estimated time for accumulating all the needed matching pieces is . . . ten years. A second possible solution is to simply forego repairing one or a few of the small holes in the wall on the downhill side of your compound that form by accident every day, allowing the water to flow through. Of course the second course of action is the only realistic one. You have an urgent problem that needs to be solved *right now*. There's no time to wait for fancy solutions. If nothing is done soon, you won't be around to benefit.

Now suppose ten years have passed. One day, quite by accident, pieces of debris that could be made into a pump fall into your compound—if you needed one. But what purpose would a pump now serve? Any extra water simply flows through holes in the wall. The need for a pump has long since passed, so you throw away the unnecessary junk.

Switching back to biology, as I argued in *Darwin's Black Box*, classic full-blown irreducible complexity effectively prohibits the development of intricate molecular machinery by mutation and selection. Yet even the simplest mIC features of comprehensively complex proteins are severely problematic. At a minimum, the time needed for

Figure 9.5. The water is rising quickly. Should the man wait for delivery of a complex pump that's on a ten-year back order from the hardware store? Or should he punch a hole in the wall to let the water drain out?

random processes to find a complex solution to a problem is vastly greater than for a simple solution. That guarantees two outcomes: (1) almost any quick, desperate yet helpful fix that involves only a single change will be selected first, even if the quick fix damages or degrades preexisting structures; and (2) the quick fixes will tend to obviate the problem—by the time a complex constructive feature saunters onto the scene, it's no longer needed.[32]

When responding to David Snoke and me, Michael Lynch wrote that, using the assumptions of his optimistic model, "adaptive multiresidue functions can evolve on time scales of a million years (or much less)."[33] Okay, much less—let's say a hundred thousand years. But, as Richard Lenski's experimental work (described in Chapter 7) shows so clearly, beneficial damaging mutations evolve on a time scale of *weeks*. That's at least a *million times* faster than the *simplest* mIC features evolving by the *fastest* route imagined. To put that in perspective, damaging mutations are like packages delivered across the country by FedEx; mutations to construct mIC features are like packages delivered by turtles.

In the real world, any possibly beneficial degradative mutations will arrive rapidly, in force, to alleviate any selective pressure on an organism—eons before the first multiresidue feature even appears on the scene (Fig. 9.5). The result is that *every* degradative change and *every* damaging single-step mutation would be tested multiple times as a solution (or as part of a solution) to whatever selective pressure a species was facing and, if helpful, would spread to fixation well before a beneficial multiresidue feature even showed up. Where Darwinian processes dominate, the biological landscape would be expected to be littered with broken but helpful genes, damaged yet beneficial systems, and degraded organisms on crutches ages before any fancy machinery was even available. That's exactly what we saw in Chapter 7 with laboratory *E. coli*, natural *Yersinia pestis*, wild polar bears, tame dog breeds, and all other organisms so far examined.

No Escape

The baleful aspects of random mutation, natural selection, and irreducible complexity are not incidental. They don't just cause difficulties that can be avoided if we're careful. Rather, they are intrinsic facets of those phenomena—flip sides of the very same coins whose positive features are so widely celebrated. Random mutation, natural selection, irreducible complexity—they supply variation, sharpen a system's function, and allow for the existence of true machinery. Yet in the same way and by the same mechanisms, they also break things, ossify a system, and greatly or indefinitely delay the appearance of a feature. The less desirable aspects aren't additional, previously hidden properties. Rather, they are the same forces working in the same ways as what Darwin's theory always claimed for them; *it's what they do.* The only radical new development is science's ability to probe life in sufficient depth, accuracy, and detail to follow crucial functional changes at the molecular level.

And it's not only what they do; it's what they've always done. A hundred years ago, a thousand years ago, a million years ago, a billion years ago—there never was a time when those natural forces were free of their downsides. There never was a golden age when only the constructive sides were working while the damaging sides were constrained, no more than there ever was a time when gravity couldn't be destructive. That's a critical point to grasp, because some puzzling features of life have been attributed to a greater latitude for Darwinian evolution in the far past than is thought to be available today. For example, over five hundred million years ago a rush of new "body plans" (think of the way basic parts of a body are arranged differently for clams, spiders, and cows) arose in the blink of a geological eye in the Cambrian period. Since the Cambrian explosion, no new innovations at such a fundamental biological level have occurred. Perhaps, some have thought, evolution just had more freedom back then—it only got bogged down more recently. The same sort of thinking is invoked with other complex biological systems that seem to have popped up in the past but don't do so today, such as gene regulatory networks, novel proteins, and even life itself.[34]

But that's wishful thinking. Throughout the vast ages of the earth, whenever genes or proteins existed, random mutation could helpfully break them, natural selection could rigidify them, irreducible complexity could radically delay any new complex features. A strong example of this has been discovered recently. Novel techniques allow the genomes of bacteria to be sequenced straight from the raw environment, without first having to be grown in a laboratory culture. After analyzing over a thousand such genomes a group announced in 2016 that the number and types of bacteria are enormously greater than had been thought. In particular, a huge group of ancient bacteria shared an interesting characteristic: "Thus far, all cells lack complete citric acid cycles and respiratory chains and most have limited or no ability to synthesize nucleotides and amino acids."[35] They seem to have lost those essential features long ago.

From the dawn of life to the present, beneficial degradation has been a constant background—there's no way to avoid it. From the beginning the Darwinian mechanism has been self-limiting, capable to an extent of eliminating or modifying preexisting molecular systems and in the process giving rise to new varieties of creatures below the biological classification level of family (described in Chapter 6), but incapable of building functionally complex molecular structures. To explain them, we must look elsewhere.

Saint Elsewhere

But where else? Although I've spent most of the book discussing it, Darwin's isn't the only theory of evolution on offer these days—"neo" or not. As detailed in Chapters 4 and 5, a substantial number of scientists, discontented with the current state of affairs, have weighed in with potential supplements or alternatives, from the neutral theory championed by Michael Lynch, to the complexity theory investigated by Stuart Kauffman, to the inclusive inheritance and niche construction theories proposed by proponents of the extended evolutionary synthesis, to the natural genetic engineering theory put forth by James Shapiro. Can one of them pick up the ball that Darwin fumbled?

The answer is a flat no. Of course, each of the proffered alternatives points to one or a few classes of phenomena that it has a reasonable shot of accounting for, at least in part. But none of them have the resources to explain the basic, functional, sophisticated molecular machinery of life. In fact, none even try to do so. Neutral theory by definition can't account for functional systems, while evo-devo and natural genetic engineering presuppose them in hopes of getting more. As for other parts of the EES, well, it's hard to see how the two-legged-goat effect would account for the elegant systems detailed in Chapter 2.

Worse than their theoretical shortcomings are their experimental ones. In Richard Lenski's fifty-thousand-generation bacterial-evolution experiment, none of the mechanisms of EES proponents were anywhere to be seen, save perhaps for the degradation of some genes by mobile genetic elements. In the *de*volution of *Yersinia pestis* or dog genomes, the speciation of polar bears or mammoths, the radiations of the African cichlid or Galápagos finch, if any of the ballyhooed alternative mechanisms of evolution played a part, it has yet to be described. As I noted in *The Edge of Evolution*, in an astronomical number of malaria cells exposed to the antibiotic chloroquine, no fancy alternative evolutionary mechanisms helped the parasite develop resistance. Only a couple of classical random point mutations in the gene for a single protein plus run-of-the-mill Darwinian natural selection were effective.

In my view, Darwin wins hands down in the contest for the best of the totally inadequate mechanisms. His theory truly accounts for the marginal or damaging changes in the machinery of life that can indeed affect important biological spandrels, from the development of antibiotic resistance to variation at the level of genus and species. But what accounts for the machinery itself? In our uniform, unbroken experience, what is the *only* explanation for the purposeful arrangement of parts? In Part IV—the next chapter—we finish our journey with that decisive question.

PART IV

Solution

CHAPTER 10

A Terrible Thing to Waste

Let's pause for a moment to recapitulate what we've learned. In the *Origin of Species* Charles Darwin argued at length for a novel yet simple idea. Competition in nature would surely yield the preferential survival of organisms whose variant biological traits best fit them and their offspring to the environment; when repeated over and over again through countless generations that might well lead to new species such as those found on the Galápagos Islands—and in the process even build up all the wonderful structures of life such as the eye. Until then nearly all thinking people had attributed complex functional traits to purposeful design.

But of course Darwin did not *show* that apparently purposeful systems could be built by natural selection acting on random variation. Rather, he just *proposed* that they might. His theory had yet to be tested at the profound depths of life. In fact, no one then even realized life had such depths. Darwin built a case with the best science available in the nineteenth century. The case was pretty strong for a few of his theory's multiple aspects, including the descent of modern organisms from earlier ones. It was extremely weak for his proposed mechanism of evolution. A major reason for its weakness is that the science of Darwin's day had no understanding of the molecular foundation of life. Only now, only within the past twenty years has science advanced sufficiently to examine life in the molecular detail necessary to rigorously test Darwin's ideas, particularly what I've termed his crucial first and last theories (that is, the presumptions that complete randomness underlies life and that repeated rounds of random mutation and natural selection can build coherent biological systems).

As we've seen throughout this book, Darwin's mechanism (as well as proposed extensions of it) fails for all but the most modest adaptations. Since even the smallest need for coordination—for even the tiniest mini–irreducibly complex feature—is a huge problem for random evolution, the Principle of Comparative Difficulty tells us why more complex structures are beyond the reach of Darwinian processes. What's more, modern research reveals that his mechanism suffers from a second, previously hidden, fatal weakness: not only are random mutation and natural selection grossly inadequate to *build* complex structures; they strongly tend to *break* them. Darwin rightly touted natural selection as relentless, as "daily and hourly scrutinising . . . every variation, even the slightest; rejecting that which is bad, preserving and adding up all that is good."[1] Yet, since the mechanism has no foresight, and since in many circumstances the random *damaging* of genes can be *helpful* to an organism, then selection "adds up" those degradative changes only in the sense that broken pieces of machinery might be added to a growing pile of junk.

Its inexorable predilection to hastily squander genetic information for short-term gain—encapsulated by the First Rule of Adaptive Evolution—guarantees that Darwin's mechanism is powerfully *de*volutionary and explains why unguided evolution is self-limiting. Ironically, random mutation and natural selection *do* help form new species and new genera, but chiefly by promoting the *loss* of genetic abilities. Over time, dwindling degradatory options fence in an evolutionary lineage, halting organismal change before it crosses the family line.

And Now for Something Completely Different

If neither Darwin's nor any other proposed physical mechanism accounts for the elegant structures of life, what does? To answer that question, this final chapter will radically shift focus from the surface level of science to its philosophical root. Failure to recognize the conspicuous explanation for life is due wholly to the explicit denial by evolutionary biology and other contemporary scholarly disciplines of the necessary foundation for any kind of knowledge—that *mind is real*.

The reality of mind will surely strike most people as something that is too obvious to need mentioning, let alone defending. But in a world where our finest institutions of higher education house learned professors who write about mental zombies (automated beings with the appearance and behavior of real people but no consciousness[2]) and who argue that our bodies go through life with their thoughts and motions determined solely by chemical reactions in our brains—well, in such a world it is the duty of us all to constantly defend the obvious. The academic ideas of nutty professors don't always stay confined to ivory towers. They sometimes seep out into the wider world with devastating results.

There is a huge scholarly literature on evolutionary biology and the philosophy of mind, many books and many more journal articles. But the best short statement of the dominant academic view of mind is by Francis Crick, codiscoverer of the DNA double helix, at the start of his 1994 book *The Astonishing Hypothesis: The Scientific Search for the Soul*:

> The Astonishing Hypothesis is that "You," your joys and your sorrows, your memories and your ambitions, your sense of

> personal identity and free will, are in fact no more than the behavior of a vast assembly of nerve cells and their associated molecules. As Lewis Carroll's Alice might have phrased it: "You're nothing but a pack of neurons."[3]

Crick's position can be more formally labeled as neo-Darwinian materialism: our minds are nothing more than our physical brains and nervous systems, shaped entirely by random mutation, natural selection, and other irrational forces. I'll let his statement stand for the entirety of the literature that shares that sentiment. In the next section we'll see that Crick's views are built on an assumption that is no longer tenable (if it ever was). Its refutation will also be the key to understanding what accounts for the structure of life. For now I will simply stipulate at the outset against Crick what almost every sentient person knows: we are truly conscious (yes, some people deny that); we have minds; we have free will; we are intelligent; and we know these things by introspection more firmly than we know any fact about the external world. Denial of any of those statements is self-refuting, like a person who denies that he or she exists. Those who declare they have no mind, are not intelligent, conscious, or free are hardly in a position to reason about any topic, let alone about the state of the mind they deny having.[4]

Yet how can we tell that another mind besides our own exists and has acted? In this final chapter we'll first explore how we recognize the effects of mind. After that we'll be fully prepared to consider the question of what accounts for the machinery of life.

A Doomed Division

How did science—the very discipline we use to understand the physical world—get to the bizarre point where some otherwise very smart people use it to deny the existence of mind? Arguably it started innocently enough. At the urging of the philosopher Francis Bacon, a contemporary of Shakespeare, four centuries ago science made a crucial decision. It would abandon the old idea of "final causes"—that is, the notion of the *purpose* of an object—which it had inherited from Aristotle. Whether the true role of, say, a waterfall or a forest is to exhibit the glory of God, supply beauty to the world, or something else couldn't be decided by an investigation of nature alone. Henceforth science would leave all such questions to philosophy and theology, restricting itself to investigating just the mechanics of nature. What a cow or mountain or star is "for" would trouble science no longer.

It seemed like a good idea at the time, and science of course has made tremendous progress since then. But such a simplistic, binary distinction was doomed from the start, because some parts of nature are very much "for" certain things and can't be understood apart from their functions. The purpose of a horse might be obscure, but the purpose of a horse's eye is not. The "function" of a stone can't be decided by science, but the function of a heart surely can.

For two hundred years the new division of labor between science and philosophy rested uneasily—until the truce was shattered by Darwin. Ignoring the tenuous peace treaty, Darwin once again addressed the question of purpose from within science itself, but this time in order to forthrightly deny there was any such thing. Apparent purpose in biology was just that—merely apparent; and in light of his theory, he wrote, "There seems to be no more design in the variability of organic beings and in the action of natural selection, than in the course which the wind blows."[5]

Biology was the last branch of science to come fully on board with the non-Aristotelian mechanistic program. As attested by many prominent evolutionary biologists, such as State University of New York's Douglas Futuyma, Darwin's lasting importance was precisely his banishment of purpose from life: "The reason that natural selection is important is that it's the central idea . . . that explains design in nature." Futuyma continued: "Darwin's (and Wallace's) concept of natural selection made this 'argument from design' completely superfluous."[6]

Despite the implicit assurances of Futuyma and many others, however, Darwin *conjectured* but certainly did not *demonstrate* that apparently purposeful systems could be built by natural selection acting on random variation (and recall that, for some aspects of life, the co-discoverer of the theory of evolution, Alfred Russel Wallace, actively argued otherwise). Those structures include the ones discussed in Chapter 2, ones featured in my earlier books, and many others. More to the point for our discussion in this chapter, by extension the structures beyond Darwinian explanation also include *brains* and *nervous systems*. A process that labors mightily to account for a simple disulfide bond is woefully unfit to account for what are likely to be the most complex, most profound structures in the universe.

Because Darwin's mechanism can't build a brain, then Francis Crick's "astonishing hypothesis" (in other words, neo-Darwinian materialism) is false. It is necessarily false, because it can't account for the very organ that Crick says is the seat of "your joys and your sorrows, your memories and your ambitions, your sense of personal identity and free will." In one stroke that refutation sweeps away the extensive literature that shares his view, because it undercuts the neo-Darwinism on which it all depends. As the prominent philosopher of mind John Searle once poignantly wrote: "We do not know how or why evolution has given us the unshakeable conviction of free will," [yet] "we cannot act except under the presupposition of freedom."[7] Searle can relax. Since, as the data recounted in this book show, random mutation and natural selection are powerless to build anything remotely as complex as a brain, then Darwinian evolution did not give us the unshakeable conviction of free will. Or the unshakeable feeling of consciousness. Or intelligence. Or mind. Something else did.

Other Minds

What is that something else? To build a foundation for the answer, the following two sections explain how we recognize intelligent activity. In order to do so I'll have to ask questions and discuss ideas that at first blush will seem downright strange, but there's no avoiding it. Whenever we ask basic questions about existence, many of the possible answers will necessarily sound odd to some people. So hold on to your hat and brace yourself for the ride.

The first question is this: We know by introspection that we ourselves have a mind, but how do we know that any other mind exists? It seems logically possible that you could be the only intelligent being around. (In fact, in its strongest form a philosophical idea called solipsism asserts that the only existing thing is the solipsist's own mind—all other people and objects are simply thoughts of that one mind.) As the eighteenth-century philosopher Thomas Reid explained, we infer the existence of other minds from their observable effects.[8] Alas, we can't read minds. We have no direct access to them. So we must use our senses to see, hear, feel, or otherwise detect what some other intelligence has done, in the same way that we use our senses to discern anything about the world outside our own minds—the same way that science investigates anything about the world.

What do we look for as a sign that another mind is present or has acted? After all, there are plenty of things in our world that seem not to have minds and lots of effects that seem random. What is it about the things that do seem to be intelligent that gives us that impression? Is the telltale sign simply that we know ourselves to be intelligent, so when we see another organism that physically resembles ourselves—another human—then we are justified in thinking that person is also intelligent?

Although that initially seems plausible, it can't be right. For one thing, if we take intelligence to mean the same thing as resemblance to ourselves, then, since no one resembles us more closely than we do, that means we ourselves would be the most intelligent thing around. Although many of us know people who do think exactly that, most of us have sufficient humility to reject a strict resemblance-to-me criterion for mind. We can see even more problems for the resemblance hypothesis from the other side. Suppose we visited an institution for the profoundly mentally challenged. Even if all the residents looked similar to us, we would be unsure if any particular person had a functioning mind unless they did something to demonstrate it. How could they do so?

A good example of what we do in fact look for to demonstrate intelligence can be seen in the 1968 movie *Planet of the Apes*. A spaceship carrying astronauts crash-lands on a planet ruled by other intelligent primates—gorillas, orangutans, chimps (all of whom, of course, speak perfect English)—while the native, humanlike creatures of the planet are unspeaking animals living in the wild. The plot contrives to have one of the astronauts (Charlton Heston) suffer a throat injury so he can't speak and be captured along with some animal-humans by an ape hunting party in a roundup. None of the ape captors can tell that the astronaut is intelligent by resemblance to themselves. Yet later, after his injury has slowly healed, in a dramatic moment while he is being harassed he yells out, "Take your stinking paws off me, you damn dirty ape!"

Immediately they knew this creature that resembled an animal of their world was actually intelligent, because he spoke their language. Although other animal-humans made sounds, the astronaut purposefully arranged sounds into a sentence that carried meaning. Thus the key is this: *because minds can choose to order whatever is within their power to manipulate, intelligence is detected by perceiving a purposeful arrangement of parts*. That is *the* way, the *only* way, that we can discern the existence of other minds and their intelligence. The "parts" that are arranged can be virtually anything: words, actions, objects, events, and so on. Yet absent a purposeful arrangement, we cannot tell that another mind exists.[9]

I said earlier that we have to recognize an arrangement of parts with our senses because we can't read minds. But even if we could read minds, we would still determine intelligence that way. Another movie illustrates that point. In the decidedly sexist 2000 romantic comedy *What Women Want*, Mel Gibson plays a character who through a silly accident gains the ability to read women's minds (which he uses for selfish purposes before being redeemed in the end). In one scene, though, two adoring female assistants smile at him, but their minds are blank—nothing to read. The scene is played for comic effect, but the point is true nonetheless. Even if we could read minds, we would only know we had encountered one if it were doing something intelligent—that is, if it were purposely arranging its thoughts.

Intelligence comes in degrees and, again, we can only determine how intelligent a mind is through its actions—through the more or less sophisticated purposeful arrangement of parts. Just one more movie example. In the 1980 Star Wars episode *The Empire Strikes Back*, the hero, Luke Skywalker, crash-lands his small spaceship on a distant world. While trying to repair it, he is irritated by a small, seemingly semi-intelligent creature he tries to shoo away. After a while, however, the creature reveals himself to be Yoda, a wise Jedi master. Yoda was sandbagging Luke, exhibiting much less than his total intelligence in order to draw him out. And, as for the case of Yoda, so for all else. We determine someone's intelligence by what they can do, by parts they can arrange.

A real-life example is the case of Jean-Dominique Bauby, who was editor of the fashion magazine *Elle*. At the age of forty-three he suffered a brain hemorrhage that left him with locked-in syndrome, a condition in which he was unable to move—save for blinking his left eye. A year later, from his hospital bed he dictated a profoundly moving, uplifting book on his experiences, *The Diving Bell and the Butterfly*, by blinking his eye in code to a transcriber. He began: "Through the frayed curtain at my window, a wan glow announces the break of day. My heels hurt, my head weighs a ton, and something like a giant invisible diving bell holds my body prisoner."[10] On what basis would a casual visitor to the hospital ascribe such acute mental powers to Bauby? As these examples demonstrate, we can tell that a mind is at least as intelligent as its actions have shown, but it might be more intelligent—perhaps much more.

It's easiest for us to see intelligence in the use of words, since that's how we most efficiently express our own minds. But it's important to recognize that language is a subset of the more general category "parts, purposeful arrangement of."[11] In speech we arrange sounds into words, words into sentences, sentences into a conversation. Yet literate people can also do the same with physical marks—writing—instead of their own voice. Correspondence, books, journals, encyclopedias, and other written artifacts are paradigms of intelligent activity.

With the category of writing, the link between the recognition of intelligent activity and the physical proximity of a candidate for possessor of the mind that arranged it is broken. In less stilted language, an author's work doesn't have to be in the same room as the author is. A book can be far away in both distance and time from the mind that composed it. If explorers in the far future unearthed a copy of the *Iliad* or *The Canterbury Tales* or *The Joy of Cooking*, then if they could translate it, they would know immediately it was the work of a mind. The lesson is that the effects of a mind can be recognized by the purposeful arrangement of parts, even when the possessor of the mind that arranged them is nowhere to be seen.

Mind Built the Machinery of Life

Mind is perceived not only in spoken or written words, but in the purposeful arrangement of anything—events, for example. Suppose a killer dispatched a victim with such care and planning that no investigator could distinguish the murder from an accident. The foulness of the deed might remain forever unrecognized. If the killer later did the same to ten victims in the same careful way, his *method* might remain undetectable. However, if all the victims had previously been scheduled to testify in the killer's upcoming drug trial, we would be certain it was murder. We would easily discern the *purpose* in the arrangement of the events.

Intelligence is also perceived in the purposeful arrangement of physical pieces. Stones placed into the shape of, say, an arrow pointing the way back to camp testify to the mind that conceived it. And, most especially in our era, intelligence is seen in the arrangement of pieces of complex machinery that are shaped to fit with each other to form a purposeful coherent whole—anything from the most advanced computers down to a humble mousetrap. In any of those arrangements we easily recognize a designing mind.

As for the rest of nature, so too for living things. The machinery of life is stunningly sophisticated, so much so that the overpowering appearance of design is acknowledged by virtually everyone, even by those who doggedly resist that conclusion. For example, on the very first page of his classic 1986 book defending Darwin's theory, *The Blind Watchmaker*, Richard Dawkins writes: "Biology is the study of complicated things that give the appearance of having been designed for a purpose."[12] Notice that's the *very definition* of biology according to Dawkins: the study of things that appear designed. What's more, he happily agrees that the appearance of design in life isn't marginal; rather, it's "overwhelmingly" strong.[13]

Richard Dawkins of course doesn't think that life was in fact designed, the appearance of which he calls an "illusion"; he thinks Darwin's mechanism did the job. So why, counterfactually, does he think it even *looks* designed? Is it perhaps for aesthetic reasons—maybe because flowers are so pretty and puppies are so cute, it seems someone must have made them that way? No, according to Dawkins life looks designed not because of aesthetic reasons, but because of *engineering* ones—life looks like the work of "an intelligent and knowledgeable engineer."[14] In other words, life looks designed exactly because of its purposeful arrangement of parts. It overwhelmingly appears like the work of a mind.

Dawkins wrote contentedly in 1986 of the overpowering appearance of design in life, because he thought he had in hand a different explanation: random mutation coupled to natural selection. But only in the past twenty years have scientific methods been developed that can probe the molecular level of biology in sufficient detail to test Darwin's mechanism. As we've seen throughout this book, random mutation and natural selection can't accomplish anything remotely like what has been ascribed to them. Consequently, the actual "illusion" is a thoroughly modern one—the illusion that Darwin's or any other proposed evolutionary mechanism can account for the elegance of life. Their supposed power was all in our heads.

Biology is suffused with a multitude of parts arranged purposefully, especially at its foundational level. Bacterial flagella, tank treads, sophisticated gene regulation, insect gears—all of those display more purpose than many of the things in our everyday lives whose design we instantly recognize. The degree of intelligence exhibited in life's physical structures is light-years beyond what we modern humans have the capacity to produce. And, as for the case of Yoda, the intelligence we perceive is the lower bound for the intelligence that the designing mind possesses. Its actual intelligence might be very much more. Although chance events certainly do occur and can leave their imprint around the far margins, from its purposeful physical structures we can *firmly* conclude that, to an overwhelming extent, life is the product of a mind.

Science Versus Reason

Despite the shock that such a statement induces in some corners of our modern culture, it's really a trivial, blatantly obvious deduction. The same conclusion of purposeful design for the surface level of biology was nearly universally shared by all thinking persons for all of recorded history until comparatively recently. Although earlier people did not have our advantages in science, they did know how to reason from a purposeful arrangement of parts. That's a lesson that modern science and other disciplines will have to relearn. The Enlightenment separation of science and purpose seemed like a good idea at the time, but it wasn't. Reason is a unity, and arbitrary divisions of reason can lead to cognitive disaster, as this and the following section will show.

Science has achieved such prominence in our modern world that we sometimes forget that it depends radically on more fundamental ways of thinking. Although it's hard to define *science*, a working definition could be something like: "The observation, identification, description, experimental investigation, and theoretical explanation of phenomena."[15] That's good enough for our purposes here, because I just want to highlight what is *not* included in the definition—that is, what is prior to science, what it stands on.

An easy example is mathematics. Math is a separate discipline from science, yet of course much of science is critically dependent on mathematical reasoning. Some historians have argued that the turning point between ancient and modern science came in the 1600s when the English physician William Harvey first used mathematical analysis to show that blood had to recirculate. He calculated that on average 540 pounds of blood was pumped by the heart each hour—far too much for it to just sink into the tissues, as the ancient Roman physician Galen had taught.[16] If it weren't for mathematical reasoning, modern science wouldn't be possible.

The same can be said for even more basic modes of thinking, such as simple logic. Deduction, induction, syllogisms, the principle of sufficient reason, and more—none of those were independently demonstrated by experiment. All of them are more basic than science, and science depends on them in order to do its work. A deeper example that's closer to the point I'm going to make is the reasonableness of believing in a real world separate from our own minds. A radical philosophical school of thought called ontological idealism held that only thought is real—the physical world is merely apparent.[17] Yet in order to investigate nature, one has to be confident there is a nature to investigate. No experiment can show it without begging the question.

Science depends on the rational belief in a world independent of our thoughts. Closely related to that basic aspect of rationality is our belief in the existence of other minds and that we can reliably detect their existence through the purposeful arrangement of parts. If we were unable to detect other minds, we would be locked into a solipsistic world where ours was the only mind we could know.

Finally, the basis of all science, of all reason, is our confidence that we ourselves have a mind. If we do not possess a real, functioning mind that can grasp the truth about nature—if we're the equivalent of a brain in a vat, fed sensory impressions by processes unrelated to truth—then we can know nothing about the world, understand nothing about reality at all.

Francis Crick's materialistic neo-Darwinian notion that we don't actually have minds sounds silly. It is quite literally absurd, and the very great majority of people go about their daily lives without entertaining such thoughts. In fact, it is hard to see how people could go about daily life if they took the view seriously. Yet ideas can have hidden implications that unfold slowly and filter down into society over time. One implication of Darwinian materialism (it wasn't "neo" yet) eventually dawned on Darwin himself. In an 1881 letter he wrote: "But then with me the horrid doubt always arises whether the convictions of man's mind, which has been developed from the mind of the lower animals, are of any value or at all trustworthy. Would any one trust in the convictions of a monkey's mind, if there are any convictions in such a mind?"[18]

Great question. If our "minds" have been formed by random mutation and natural selection, which aim only for survival and reproduction, why should we think they give us access to the truth? The modern philosopher Patricia Churchland takes the bull by the horns, declaring that a more powerful brain "is advantageous [only] so long as it . . . enhances the organism's chances of survival. *Truth, whatever that is, definitely takes the hindmost*" (emphasis added).[19] Churchland and her philosopher husband, Paul, are exponents of something called *eliminative materialism*, which among other things holds that "common-sense mental states, such as beliefs and desires, do not exist."[20] Other philosophers have denied the very existence of consciousness. Borrowing the idea from Richard Dawkins and running with it, the psychologist Susan Blackmore thinks all minds, including her own, are composed of "memes"—little idea fragments that reproduce autonomously in brains. All such views share a common starting point—materialism. Yet, as the late philosopher of science Paul Feyerabend thought, "practically any version of materialism would severely undermine common-sense psychology."[21]

Any version of materialism undermines common sense. And "common sense" includes the notions that you have a real mind and can reason and make choices. If materialism is true—if all that exists is the matter and energy studied in ordinary physics classes—then there is no such thing as a real mind. Confronted with that dilemma, there are two choices: either affirm materialism and deny your own mind, or affirm your mind and deny materialism. As a matter of necessity, in daily life everyone acts as if they have a mind. But in public life, when writing for or speaking to other people who are aware of these issues, the sociological pressure on academics is to act as if it were not true: that humans have no mind and that our thoughts are the outplay solely of physical forces. That can lead to decidedly strange results, as the following section shows.

The Consequences of Spurning Reason

Without the fundamental underpinnings of reason—without mind and the ability to recognize other minds—science itself eventually goes off the rails, plunging deeper and deeper into irrationality. A splendid example is from Oxford University philosopher Nick Bostrom, whose 2002 book *Anthropic Bias: Observation Selection Effects in Science and Philosophy* wondered how the fine-tuning of the universe (that is, its remarkable fitness for human life) might be explained. There were two chief possibilities: "the design hypothesis and the ensemble hypothesis."[22] He spent very little space discussing design. Instead, he concentrated almost exclusively on the *ensemble hypothesis*—the idea that there are extremely many universes within a multiverse, perhaps even an infinite number, and that the physical laws and constants can vary between universes. (This is the same idea that Eugene Koonin invokes to account for the origin of life in our universe, as discussed in Chapter 4.) Since our universe contains life, the argument goes, it necessarily also has to have laws that are compatible with life.

But there's a big problem. As Bostrom explains, if the number of universes is infinite, then quantum physics seems to indicate that an infinite number of brains could just pop into existence—even in non-fine-tuned universes—already containing a host of false thoughts about their history and surroundings: "It isn't true that we couldn't have observed a universe that wasn't fine-tuned for life. For even 'uninhabitable' universes can contain the odd, spontaneously materialized 'freak observer.' . . . It is even logically consistent with all our evidence that *we* are such freak observers" (emphasis in original).[23] What would the beliefs about science of such a brain be worth? Nothing, of course. The lesson is: abandon a facet of rationality—the ability to recognize the work of other minds—and irrationality rushes in.

In a subsequent paper Bostrom upped the ante. Like some academic version of *The Matrix*, he argued that we are probably living in a computer simulation: "One thing that later generations might do with their super-powerful computers is run detailed simulations of their forebears. . . . We would be rational to think that we are likely to be among the simulated minds rather than among the original biological ones."[24] An ordinary person might think that was just an idea for a science-fiction story, yet it's taken seriously in academia, including by scientists.

The American Museum of Natural History recently hosted an academic conference on the topic "Is the Universe a Simulation?"[25] Some of the physicists participating in the conference talked blithely about what evidence they would look for to confirm the oxymoronic idea that reality is a simulation. But New York University philosopher David Chalmers came closer to articulating the crucial point: "You're not going to get proof that we're not in a simulation, because any evidence that we get could be simulated."[26] That's not the half of it. He could as well have asked why in such a scenario we think there are any laws of physics at all or what reality even means. One can only investigate a notion that undercuts reason to the extent you don't take it seriously. To the extent you do, you're paralyzed.

The event was chaired by the science-popularizing astrophysicist Neil deGrasse Tyson, who opined of the idea, "I think the likelihood may be very high."[27] Let that sink in. A major figure representing science to the public thinks that our world is probably a simulation being run in a computer somewhere. What effect will that have on young people—not only on those thinking about careers in science, but on future voters who have to decide critical issues concerning our environment? It surely can't help kids *or* adults to be told that reality isn't real and that at best science is only investigating a simulation.

Historians have argued that science first took root only in Western culture because it expected nature to be rational, to be understandable.[28] No historical examples yet bear on the question of whether science can survive if a culture once again embraces irrationality. It would be foolish to put that question to the test.

Castles in the Air

Most people have too much common sense to swallow the idea that reality is a mere computer simulation. Yet the same underlying concept—the notion that we can't know the real world, that our own minds are products of forces that aren't aiming for the truth—is found a lot closer to home in what's now termed "evolutionary psychology." Roughly, that's the idea echoed in the previous quote by Patricia Churchland, that our minds have been formed by evolutionary forces—by Darwin's mechanism—that built us merely to survive, not to understand or act rationally.

That then unnamed notion burst into the American public consciousness in the 1924 murder trial of Nathan Leopold and Richard Loeb for the thrill killing of a fourteen-year-old boy, Robert Franks. Leopold and Loeb's defense attorney (Clarence Darrow, who a year later would square off over evolution with William Jennings Bryan in the Scopes monkey trial) raised a novel defense, that the forces of evolution made them do it: "Science has been at work . . . and intelligent people now know that every human being is the product of the endless heredity back of him and the infinite environment around him."[29] In other words, science has been at work to show we have no minds, so killers can't be held responsible for their acts. But if that's right, what is the difference between a mind programmed in a universal computer simulation and a mind programmed by evolution? In either case there really is no mind, just a program.

The Leopold-Loeb trial was long ago. The neo-Darwinian materialistic notion that our minds are only what they have been selected to be has since grown much stronger. One case in point is the 2000 book published by MIT Press, *A Natural History of Rape: Biological Bases of Sexual Coercion*, which argues that such violent behavior has been favored in some circumstances by natural selection. It's easy to imagine Leopold and Loeb's lawyer at trial holding up a book called *A Natural History of Thrill Killing* if it had been available in 1924. The effect isn't limited to sensationalistic topics. That's the same intellectual river from which flows the stream of books cited in Chapter 9 with titles such as *A Darwinian Left: Politics, Evolution and Cooperation* and *Literary Darwinism: Evolution, Human Nature, and Literature.* The irrational notion that we don't have minds, that we are the sum of the Darwinian evolutionary forces that supposedly produced humanity, is deeply embedded in our culture.

Yet as the book titles themselves readily show, almost all modern materialism rests on a Darwinian foundation, so it's all built on a cloud. It's astonishing to think of all the work that's been premised on what even in its heyday was at most a promising hypothesis. There never was any hard evidence that Darwin could build the coherent machinery of life, let alone our brains, let alone our thoughts. It can't be repeated too often that it's only been twenty years since science gained the ability to test Darwin's mechanism at the critical molecular level of life—the level that carries hereditary information, the level that contains the most sophisticated machinery. Now that it has been tested, we understand why an unintelligent process that can barely manage to put together a hook-and-eye latch can't make a supercomputer far surpassing anything humanity has yet built. The desperate need to toss complex machinery overboard to save a sinking evolutionary ship won't somehow build the machinery in the first place. Darwin's mechanism can't begin to make a comparatively simple bacterial flagellum, let alone the human brain. Thus all of the intellectual work built on that vaporous foundation falls with it.

Just as design reaches deep into physical life, so it reaches deep into mind. There's no reason to think that, even starting with a working brain, random mutation and natural selection could coherently change some mental computer program, even if such a thing controlled parts of the human brain. What series of random changes to a brain program—that is, what poison-pill mutations to nervous system genes—would build some consistent thought? Thus the whole enterprise of evolutionary psychology, built on the entirely fictional constructive power of Darwin's mechanism, is misguided.

In mathematics, division of a number by zero is undefined—that is, it has no meaning within the system. If people ignore the rules and divide by zero anyway, they can contrive to get any result they choose. As advantageous as that may sound to beginning students, the result bears no relationship to reality. Similarly, reason is all of a piece. One can't accept some of it and ignore other parts. It's a package deal. If one accepts the principles of deduction and induction, but spurns the truth that the world really exists, you can get any result you want, but you've lost the connection to reality. Worse, if you lose confidence that you have a mind that can lead you to the truth, you become trapped in a world that can't be known.

The illegitimate mathematical division by zero can be done by mistake in long, complex calculations. A person might only realize something's wrong after strange results come in. The denial of reason can be done by mistake too in seemingly sensible ideas whose outlandish implications take a long time to surface. But when they do surface, don't let yourself be talked out of your mind by enthusiasts for the notion. A basic aspect of reason is our ability to recognize the existence of other minds. If we lose confidence that we can perceive the work of another mind through the purposeful arrangement of parts, we are stuck in a solipsistic universe, perhaps even imagining ourselves as brains that have popped out of the void with false thoughts or as existing in some weird computer simulation.

The denial of the reality of mind is a worse calamity for science and the society it informs than ontological idealism, which denied the reality of matter. As the history and philosophy of science has shown in the last 150 years, when we lose the ability to recognize the work of another mind in the powerfully purposeful arrangements of nature, we lose the ability to recognize even our own minds.

A Classic Problem

In the past few sections of this chapter we've seen that the action of a mind is discerned *uniquely* and *explicitly* in the purposeful arrangement of parts. In earlier chapters we saw that the machinery of the very foundation of life is itself overwhelmingly arranged for purposes. Thus we can once again confidently conclude that life is what most people over the ages have taken it to be—a product of mind. That single conclusion, however, does not of course mean all related problems have been solved. Rather, it only means that we can again be secure in *our own* rationality—in *our own* minds—and begin the task of addressing them. In this section we'll look at one classic problem: How do a physical body and a nonphysical mind interact?

In his 2004 book *Mind: A Brief Introduction* John Searle mentioned the classic problem, plus two distractions: "The failures of dualism and the success of the physical sciences, together, give us the impression that, somehow or other, we must be able to give an account of all there is to be said about the real world in completely materialist terms. The existence of some irreducible mental phenomena does not fit in and seems intellectually repulsive."[30]

Let's start by dismissing the two distractions. First, as the saying goes, there's no accounting for taste. So what seems "intellectually repulsive" to one person might have a lot more to do with social group than with reality. Second, while the physical sciences have indeed had tremendous success in many areas, as we've seen in this book those areas conspicuously do not include accounting for the origin of the complex structures of life. To the extent philosophers thought otherwise, they were simply drawn in by the presumptions and enthusiasms of Darwinian evolutionary biologists.

Now for the classic problem. The seventeenth-century French philosopher René Descartes (famous for his saying *Cogito, ergo sum*, "I think, therefore I am") introduced the notion dubbed *Cartesian dualism* (henceforth just *dualism*), which says that humans are composed of two completely separate substances, a physical body and an immaterial mind. Aristotle had previously reasoned that humans were just one substance, but that the substance is an amalgam of matter and form (for example, the form, or shape, of a bird can be imposed on the matter of clay to make a statue). That notion was called *hylomorphism*.

Descartes's new idea came with a new problem: How does the immaterial mind interact with the material body? Aristotle hadn't thought there was any problem in the first place, because in his thinking it was simply a power of the form to affect the matter; that is, mind and body were all of a piece. Descartes's dualism was ridiculed as the "ghost in the machine"—how could an ethereal entity move a physical one, or vice versa?[31] As John Searle wrote, in the absence of any good answer and in light of the success of physical science, dualism was discredited, hylomorphism forgotten, and the problem of the ghost in the machine dodged by the expedient of eliminating the ghost, leaving the bare machine. From then on, the mind was assumed to be just another physical phenomenon, no different in kind than digestion.

Frankly, that's crazy. I have no answer to the problem of how the mind affects the body or the reverse, but denying your mind because you can't solve a problem is like cutting off your head to cure a headache. Whatever difficulties dualism, hylomorphism, or some other proposed explanation may have, they pale in comparison to denying mind. When you make that move, no more arguments are left, because—to the extent you are consistent—there is no more mind to reason about them. In the same way, scientists who embrace ontological idealism are finished, because there is no more nature for them to investigate.

What's more, the usual difficulties listed for mind–body interaction strike me as ranging from superable to trivial. One consideration that's always mentioned is that for the mental to affect the physical would contradict the principle of determinism—that the laws of nature are inviolable and alone determine the behavior of physical objects in a billiard-ball fashion. But science already agrees that determinism is false. At the most basic level of matter, the quantum level, events are understood by most physicists to be physically uncaused. Perhaps there are nonphysical events that can affect quantum ones in a purposeful way, in turn affecting the brain.

Another frequently stated worry is that such events might violate the principle of the conservation of energy. A skeptic might reply, so what? How much energy need be involved anyway? Whatever it might be, it would likely contribute far less to global warming than other processes in nature. A further complaint is that neuroscience hasn't been able to identify events in the brain that have been affected by the mind. Yet, as discussed in the first chapter, no branch of science can currently account for even purely physical complex events. Why think mind–body interactions should be any easier?

None of the above suggestions need turn out to be true to justify the existence of mind. Perhaps some other notion will solve the mind–body problem; perhaps not. Maybe we'll never have an explanation for mind. But even if no explanation turns up, that's no reason at all to deny the existence of the faculty through which we know things in the first place. If science never finds an ultimate explanation for matter, should its existence be denied? Should all scientists become ontological idealists? Yet that silly suggestion is less absurd than denying your own mind.

I should add that I am not at all saying that material things can't influence the mind, including alcohol, drugs, genes, environment, organic mental diseases, and more or that science can't contribute tremendously to understanding their effects. The point is that, although material things do influence the mind, they do not constitute it.

Whose Mind?

One big question of course is: Just who is this mind that's behind life? As I've explained in previous books, the question of the identity of a designer can be a much tougher question to answer than the question of whether something was designed. As a quick example, the first European explorers immediately knew that the statues on Easter Island were purposely made many years before anyone had a good idea who might have built them or how. In a science-fiction example, if space explorers landed on a deserted planet that contained sophisticated machinery, they could easily know that there had been a mind responsible for the machines even if they couldn't figure out its identity. We're in a similar position, except that instead of the far reaches of space we're exploring the depths of the cell. The clearly designed machinery is there to see, but who designed it?

Most people, including myself, are theists and will naturally tend to ascribe the design to God. But I want to emphasize here that the idea of teleology behind nature is expansive; plenty of intellectual room remains for people of widely varying philosophical inclinations. For example, several years ago the eminent New York University philosopher of mind Thomas Nagel wrote a book, *Mind and Cosmos*, which has the most trenchant subtitle I have ever seen: *Why the Materialist Neo-Darwinian Conception of Nature Is Almost Certainly*

False. Nagel is himself a committed atheist. Nonetheless, in his recent book he argues that science will eventually have to deal with the reality of mind, including within nature itself. In fact, he thinks that mind is an intrinsic part of nature: "My guiding conviction is that mind is not just an afterthought or an accident or an add-on, but a basic aspect of nature."[32]

Such a view has a fine intellectual pedigree. Aristotle himself viewed nature as containing intrinsic teleology, built-in purpose. And something similar seems to be implied in the natural genetic engineering theory advocated by James Shapiro. Other thinkers may have other ideas about intelligence. For example, the University of Toronto philosopher of mathematics James Robert Brown, author of *Smoke and Mirrors: How Science Reflects Reality*, affirms that immaterial *reasons* can be real causes of human *actions* in science and elsewhere. That's because, although he is an atheist, he isn't a materialist. Like some mathematicians, he believes in a platonic realm beyond space and time where concepts such as "triangle" and "magnetic field" actually exist. As the late Paul Feyerabend thought, it's not atheism that destroys mind, but strict materialism.

Public Understanding

In 2004 Richard Dawkins wrote *The Ancestor's Tale*, which undertook to explain evolution to a general audience.[33] The title is a play on Geoffrey Chaucer's fourteenth-century classic *The Canterbury Tales*, in which pilgrims regale each other with stories to while away the time on their trip. Instead of religious pilgrims, in Dawkins's book various animals travel along and meet up with their ancestors, starting from humans and proceeding down to bacteria. Although cute, the structure is awkward because more complex creatures appear first, followed by simpler ones. It's like a math textbook that starts with calculus, follows with long division, and ends with addition facts. Yet Dawkins chose the structure deliberately, so that humans wouldn't come last, which he worried would lead some readers to think we humans are the goal of evolution. Dawkins knows in his bones that evolution has no goal.

Early in the long book he comments offhandedly on why humans are nothing special. We have unusual abilities, sure, but so do all other creatures. Because humans have large brains, we egotistically assume that those must be the pinnacle of life. However, he writes, "a historically minded swift, understandably proud of flight as self-evidently the premier accomplishment of life, will regard swiftkind . . . as the acme of evolutionary progress." Similarly:

> If elephants could write history they might portray tapirs, elephant shrews, elephant seals and proboscis monkeys as tentative beginners along the main trunk road of evolution, taking the first fumbling steps, but each—for some reason—never quite making it: so near yet so far. Elephant astronomers might wonder whether, on some other world, there exist alien life forms that have crossed the nasal rubicon and taken the final leap to full proboscitude.[34]

"Nasal rubicon"—funny. I love nose jokes. Yet, although he just tossed it off, Dawkins isn't joking when he says humans are unexceptional. In fact, he treats it so seriously that he distorts the structure of the entire book to avoid giving any privileged position to humans. So let's return the favor and give his comments some serious *thought*. Let's ask, what exactly would elephants use to think about astronomy—their trunks? With what would they conceptualize history—their outsized ears? How would a swift "regard" anything at all—with its wings? If swifts could regard anything, they would necessarily regard their very ability to regard as the pinnacle of life. If elephants could write history, it would be a history of *ideas*. They would marvel much more at the immensity of their mental universe than at the size of their trunks.

Contrary to Richard Dawkins, the ability to reason is indeed the greatest possible power of life. The only greater gift would be the ability to reason better. The precedence of thought is not due to human arrogance; rather, it's because reasoning is a requirement for *understanding.* Yet, with sad irony, until his retirement Richard Dawkins was Oxford University's first Charles Simonyi Professor of the Public Understanding of Science.[35] His very job title presupposed the ability of humans to grasp conceptual truths about nature. A chief activity of the occupant of the chair is to explain science to the public so that they can *understand*. Yet Dawkins denigrates reason and understanding, fearful that the public will think they're something special.

Richard Dawkins is only the most visible spokesman for a view that is widely held in science and academia in general, that humans are nothing special, that what we call our "minds" are as much the product of irrational evolutionary forces as elephant trunks and bird wings. The view was implicit in Darwin's theory from the start, when he proposed to break the uncomfortable peace between science and philosophy by eliminating purpose from life. Yet if an overwhelmingly purposeful arrangement of parts can be explained as due to something other than mind, then as night follows day we lose the ability to recognize our own minds. Over time the buried implication that mind is an illusion rose to the surface and began to spread in our culture. If it's taken seriously, then there can be no professors of the "understanding" of anything—including science.

For its own good as well as the public's, science needs to *officially* reject such a view. Science is built on a rational foundation that includes mathematics, logic, the reality of nature, and the reality of other minds. Throughout history there have always been radical skeptics who denied one or more of those pillars of reason, but our modern age is the only time when the denial has become widespread within science itself. The result is worse than if ontological idealism—the denial of nature—became a majority view; at least in that case a person could still think. Just as the reality of nature is affirmed by science, so must the reality of mind be positively affirmed.

It turns out that the Enlightenment separation of science from mindful purpose could never work out well. Both aim at truth; both are required for knowledge. During their four hundred years apart science and purpose have had their own experiences and each grown in many ways, but they have also bumped up against their own limitations. It's best now to view that time not as divorce, but as just a trial separation. Happily, the couple are discovering that they can't live apart and that their reunion is long overdue. Science and purpose were made for each other.

The reunion of science and purpose should come easily, because the chief problem that divided them—neo-Darwinian materialism—has dissipated. Neo-Darwinian materialism is false, because the assumptions of neo-Darwinism are largely false. Random mutation and natural selection cannot build a brain or even coherently modify one. For those such as Richard Dawkins who fret about human pride, there is at least one consolation: there's no reason to think that bird wings or elephant trunks are the product of chance either. Those too were intended. Those and much more are all the products of an intelligence. Rather than some cosmic accident, thanks to the dazzling advance of science, those of the *public* who agree they have minds can now *understand* that nature is designed down to an intricate level of detail.

And that's a happy thought to think, because mind is a terrible thing to waste.

APPENDIX

Clarifying Perspective

The pace of modern science is astonishing. The big bang theory has rocketed from a dreamy speculation in the mid-twentieth century to a virtual certainty. Physicists can now account for many properties of the universe back to slight fractions of a second after its start. In biology, the elegant double-helical shape of DNA was first elucidated less than seventy years ago. Now whole genomes of exotic creatures are sequenced so routinely that it rarely makes news.

Yet in one area science has hit a brick wall. How could a mindless mechanism like the one Darwin proposed build the intensely purposeful systems found in life? The palpable restlessness among evolutionary biologists who think deeply about the matter (described in Chapters 4 and 5) is one key sign of the problem's intractability. Another even more telling one is the feeble substantive (if not verbal) response to a decades-long public challenge to Darwin's mechanism.

In 1996 *Darwin's Black Box: The Biochemical Challenge to Evolution* argued that the irreducibly complex biochemical systems discovered in the cell were poor fits with Darwin's theory. Machines that need multiple parts to work (exemplified by a mousetrap) cannot plausibly be made in the gradual way that Darwin insisted on. That's because incomplete intermediates either don't work at all or at the very least don't work as they would need to for the functionality of the final system. So natural selection either would have nothing to select (because the system wasn't working) or would select parts for a different purpose than needed in the end. Either way random mutation and natural selection would be grossly inadequate means for producing molecular machinery. On the other hand, because we humans detect intentional activity by observing a purposeful arrangement of parts (such as is found in abundance in machinery), the book proposed that at least some biochemical systems had been purposely designed by an intelligent agent.

The book set off an uproar—scathing editorials and court trials as well as denunciations by scientific societies, national governments, and even a committee of the Council of Europe.[1] In retrospect I don't think people were upset by the criticism of Darwin's theory or the concept of irreducible complexity nearly as much as they were by the explicit proposal of intelligent design. For a variety of reasons many scientists and others are viscerally opposed in principle to a conclusion of design for life, and some are spurred to action by it.

From my perspective, one of the most salutary effects of *Darwin's Black Box* was that it goaded some very smart scientists who were intensely opposed to its conclusions into trying to prove them wrong. The worst kind of scientific theory is one that floats along as an assumption, allowing too many people to forget that it's essentially untested. That's the way I viewed Darwin's theory, and I was delighted by the prospect that it would finally be put through the experimental wringer, to either stand or fall. Over the years some revealing studies pertaining to the mechanism of evolution were indeed published. Of course, since the topic is so exquisitely controversial, the studies were initially accompanied by a lot of spin. However, a distance of twenty years affords clarifying perspective.

This Appendix will focus on several of the most prominent scientific responses—published in peer-reviewed journals, books, and other places where academicians send their formal work—to the arguments in *Darwin's Black Box*. With this current book as background, let's see how much progress has been made in accounting for huge irreducibly complex systems by the same theory that even today struggles to account for a simple disulfide bond.

Poster Child

Phrases like "irreducible complexity" and "purposeful arrangement of parts" strike the average person as pretty obscure, at best evoking hazy, abstract images. To drive home the concepts, a writer like myself has to connect them to some everyday example. That's where the mousetrap comes in. It's so familiar to most people, so ordinary, that everyone feels pretty confident in drawing conclusions based on it. When it's pointed out that the trap won't work without all its parts and that trying to build a trap gradually à la Darwin is bound to fail if the trap has to work at each step, even people without the slightest interest in biochemistry understand. The mental image of happy little mice frolicking on an incomplete, ineffectual trap highlights the absurdity. The very effectiveness of the mousetrap in illustrating irreducible complexity made it an early target of attack by Darwin's defenders (briefly discussed in Chapter 9).

But eventually the argument has to move to the molecular level of life, where the large majority of the public is decidedly less confident. To show that the concept also applies there, a great visual illustration of an irreducibly complex molecular machine would be very helpful. In an early chapter of *Darwin's Black Box* I concentrated on two such machines: the eukaryotic cilium and the bacterial flagellum. A detailed drawing of the bacterial flagellum appeared in a popular biochemistry textbook, so I obtained permission to use it as the frontispiece—the first image readers would see.

It did the trick. Here's one typical anecdote: A colleague of mine was trying to explain the twin concepts of irreducible complexity and intelligent design to his engineer dad, who wasn't getting it. My friend pulled out a copy of the flagellum drawing. Immediately his dad whispered, "Oh . . . I see the problem."

The flagellum (Fig. A.1) is quite literally an outboard motor that bacteria use to swim. It has a number of conceptually distinct parts—a motor, stator, drive shaft, bushing materials, and more—totaling dozens of different proteins. But of course that terse description comes nowhere near doing justice to the machine's complexity. (Even the drawing, which gives the impression of some space-age contraption, falls far short, because it depicts each very complex part as a simple geometric shape. This is done for the sake of improving students' comprehension, but ironically makes it appear much simpler than it really is.) Each of the flagellum's proteins is itself intensely, comprehensively complex. What's more, unlike outboard motors assembled by humans who know exactly how to arrange the parts, machinery in the cellular world has to automatically assemble itself. As I described in *The Edge of Evolution*, the system for assembling the flagellum is both elegant and exceedingly complex. So not only is the flagellum itself irreducible, but so is its assembly system. The assembly process and the flagellum together constitute irreducible complexity piled on irreducible complexity.

Because of its powerful visual impact, the bacterial flagellum quickly became the poster child for irreducible complexity and intelligent design, making it the preferred target of Darwin's modern champions. They reasoned that if something as apparently purposeful as the flagellum could be shown to have been built by random mutation and natural selection, well, then so could pretty much anything else. Defeat the flagellum, and irreducible complexity will fall with it.

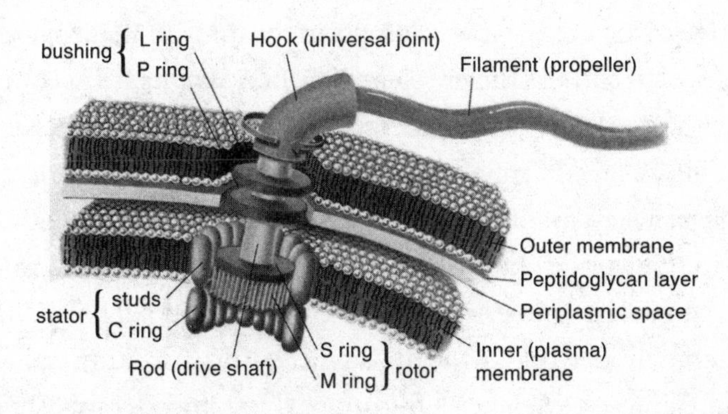

Figure A.1. The bacterial flagellum.

Easier said than done. In 1996 I showed that, despite thousands of papers in journals investigating how that fascinating and medically important molecular machine *worked*, there were no papers at all that tested how the bacterial flagellum might have *arisen* by a Darwinian process. The scientific literature was absolutely barren on the topic. Something about the flagellum made evolutionary biologists remarkably shy about even attempting to take on the challenge—now what might that be? Yet with the publication of the book as a spur, the ensuing decades of time to work on the problem, and the marvelous advances in scientific capabilities in the last twenty years, might the situation now be different?

As we'll see in the next three sections, the answer is a resounding no. Nothing at all has changed.

Comparing Flagellum Sequences

Twenty years on, there has been a grand total of zero serious attempts to show how the elegant molecular machine might have been produced by random processes and natural selection. Nonetheless, it's instructive to look at the few attempts in the interim that have at least been *claimed* to address the problem. The first serious try at coming to terms with the flagellum wasn't published until 2006, a decade after *Darwin's Black Box.* It appeared in the prestigious journal *Nature Reviews Microbiology*, an offshoot of *Nature*, the most prominent science journal in the world. The authors were Mark Pallen, a noted microbiologist then at the University of Birmingham, and Nicholas Matzke, who at the time worked for an advocacy organization called the National Center for Science Education. (Despite its comprehensive-sounding name, the NCSE's specific mission is to aggressively defend Darwinian evolution wherever it is challenged.) Under the uncertain section title "An Experimental Research Programme?" the authors candidly admitted that no research had been done on flagellum evolution until that point: "In recent years, flagellar biologists have made astonishing progress in understanding the structure, function and regulation of bacterial flagella. . . . However, the flagellar research community has scarcely begun to consider how these systems have evolved."[2]

Now, as recounted in Chapter 3 of this book, the ambiguous term "evolution" causes no end of confusion. Ernst Mayr pointed out there were no fewer than five separate ideas all wrapped up in what is termed, in the singular, Darwin's *theory* of evolution; conflating those ideas has derailed even the most distinguished of thinkers. Most important for our purposes, the concept of common descent has to be kept separate from Darwin's proposed mechanism of evolution. The fact that some organism or gene or protein may have descended from an earlier one doesn't tell us how that led to any particular structure—no more than the mere fact that a former typewriter manufacturer like IBM now makes advanced computers tells us how that led to a product shift.

Darwin's contemporaries all immediately accepted common descent, but virtually none embraced his proposed mechanism. Similarly, modern researchers like those discussed in Chapters 4 and

5 don't question common descent, but they are skeptical of neo-Darwinism. They propose new ideas such as natural genetic engineering or complexity theory explicitly to try to account for *what drove* the origin of complex functional systems. Random mutation and natural selection were the classical Darwinian answer to the same question. To evaluate his proposal, a researcher must actually test whether those twin prongs can do the job.

Pallen and Matzke's paper didn't even try to test Darwin's mechanism. Instead, as no end of studies do in these days of massive genome sequencing, the authors simply compared sequences of flagellar proteins from various bacteria (available in public databases) to look for relationships. They showed that some different kinds of flagellar proteins had some sequence similarity to each other, which supports the idea that they came from an ancestral gene. The authors also wrote that some other flagellar proteins were somewhat similar to nonflagellar bacterial proteins, which again supports the reasonable hypothesis that they are related by descent. But even if descent is correct, the authors didn't even try to test whether random mutation and natural selection were up to the massive job of drastically refashioning and arranging those proteins for a complex new role.

Like many others before and after them, Pallen and Matzke carelessly confused evidence for common descent with evidence for Darwin's mechanism. Even worse, they relied heavily on the same dubious nineteenth-century *theological* assumption that Ernst Mayr recounted in *What Evolution Is* (discussed in Chapter 3).[3] In the course of their discussion they pointed out that, although all flagella share a core of several dozen proteins, there are also many variations for different kinds of bacteria. Some flagella are thicker, others thinner; some use a gradient of acid as a power source, others use sodium ions; some rotate outside the cell, others (such as the spirochetes mentioned in Chapter 2) actually spin inside the cell.[4] So what conclusion did the authors draw from nature's lavish bounty?

> One is faced with two options: either there were thousands or even millions of individual creation events, which strains Occam's razor [that is, the notion that a simpler explanation should be preferred to a more complex one] to breaking point [*sic*], or one has to accept that all the highly diverse contemporary flagellar systems have evolved from a common ancestor.[5]

Since Pallen and Matzke confuse simple common descent with Darwin's mechanism, they clearly mean that the many alternate forms must have arisen from a common ancestor by random mutation and selection—because *God wouldn't have done it that way.* You see, a designer wouldn't have planned a lot of elegant variations on a common theme—everybody who's anybody knows that. To quote Ernst Mayr, "It seemed quite unworthy of the creator to believe that he personally arranged every detail in the traits and life cycles of every individual down to the lowest organism."[6]

What seems quite unworthy to me is the spectacle of scientists basing their conclusions almost completely on a sort of reverse theology. What God would or would not do is not within the competency of science to inquire. What is within that competency is to investigate whether random processes culled by selection could lead to *any* sort of flagellum or even to *transitions* between any nontrivial variants.[7] As we saw in Chapter 6, after millions of years a very wide variety of organisms—Darwin's finches, African cichlids, Hawaiian fruit flies, lobelias, and more—all evolved plenty of minor changes, but stalled before the classification level of families. If millions of years of such intense selection on finches as documented by Peter and Rosemary Grant can't produce anything other than a finch, then what reason besides bad theology is there to suppose it could produce significant new variations on a preexisting flagellum? Occam's razor cuts both ways.

It's well within science's competency to experimentally test evolution, as we've seen in previous chapters on the work of Richard

Lenski, Joseph Thornton, and others. Alas, no researcher—including the authors themselves—has been willing to throw valuable time down a rathole. That largely polemical paper was the only one Matzke has published on the evolution of the flagellum. Other than a couple of near contemporaneous commentaries mainly on other people's sequence work, it was the last for Pallen too.[8] And, over twenty years down the road, no one else has joined the forlorn cause.

Now Is the Time for All Good Men

A year after Pallen and Matzke's work was published, a paper appeared in the very prestigious journal *Proceedings of the National Academy of Sciences USA* bearing the juicy title, "Stepwise Formation of the Bacterial Flagellar System." The authors, Renyi Liu and Howard Ochman, then of the University of Arizona, made a startling claim: the two dozen "core" flagella proteins (that is, the proteins that are found in all kinds of diverse bacterial flagella) all arose from a *single* prodigious precursor protein by duplication and diversification of a single primordial gene. As Liu and Ochman saw it, the gene and its duplicates first produced proteins that formed the inner parts of the flagellum. Further duplication resulted in proteins to form the middle and outer parts. Then, its mission accomplished, the prodigious gene apparently rested, never to form any nonflagellar protein.

The paper made a splash. The news blog of the very prestigious journal *Science* reported the results and asked a few big names for comment. The eminent Michael Lynch (discussed in Chapter 4) remarked that "complexity builds out of simplicity, and [the work of Liu and Ochman] is a well-documented argument for how that can happen." Brown University cell biologist Kenneth Miller chimed in, "The researchers clearly show these genes were derived from one another through gene duplication."[9]

"Clearly." "Well-documented." It seemed that finally the long-standing challenge to Darwin's theory and irritating symbol of intelligent design had fallen—and so satisfyingly soon after the highly publicized trial in which it had played a starring role.[10] Yet the euphoria was short-lived. The same day that the paper was posted on the website of the *Proceedings of the National Academy of Sciences USA*, Nicholas Matzke (coauthor of the paper discussed in the previous section) was scoffing at its claims. In a blog post entitled "Flagellum Evolution Paper Exhibits Canine Qualities" (that is, it's a dog), he snorted that much of the paper "ranges from dubious to just irremediably wrong."[11] In a later commentary the eminent evolutionary biologist W. Ford Doolittle, of Dalhousie University, and a colleague, Olga Zhaxybayeva, politely termed the paper "problematic."[12]

Since its publication in 2006, little has been heard of the paper. In the past ten years the authors' thesis hasn't been explored further, either by other researchers or by the authors themselves. So how is it that such a questionable study not only was published by one of the world's leading science journals, but was eagerly publicized by the websites of other leading journals and ballyhooed by scientists who should have known better?

That is of course impossible to answer with certainty. But I can sure make an educated guess. Something. Had. To. Be. Done. A big clue is that, at a time when intelligent design was much in the news, roiling the scientific community, Doolittle and Zhaxybayeva chose to entitle their commentary "Reducible Complexity." Although I think their views are based on a misinformed caricature,[13] many researchers regarded intelligent design as less scientific than the world of Harry Potter and were alarmed by its growing popularity within some segments of society. Framing Liu and Ochman's work in terms of "a war against unreason," Doolittle and Zhaxybayeva echoed the fears of many other scientists: "Arguments about whether a flagellum could have been cobbled together, step by step, . . . figured large in the 2005 Dover, Pennsylvania, trial over the teaching of ID as science in public schools. . . . It is important that we scrutinize [Liu and Ochman's] arguments with special care, because they are likely to be under contention at the next trial."[14]

Positive down to their toes that Darwin got it right and frightened by the advance of a view they abhorred, selfless Darwinian biologists and like-minded folks contributed whatever they could to stop it. Researchers such as Liu and Ochman produced the best pertinent studies they could manage. Editors of prestigious journals gave the studies the greatest possible visibility. Science websites broadcast the news as widely as possible, certified by the enthusiastic endorsements of notable scientists. The concerted effort worked pretty well at casting a shadow over design in some people's minds. It failed miserably at explaining the flagellum. Twenty years after *Darwin's Black Box*, the stunningly complex molecular machine is no closer to receiving a Darwinian account.

The Sole Experiment

No experimental work (other than sequence comparisons) has *intentionally* been done to explain how the bacterial flagellum might evolve. But one study did so inadvertently, and it got the most revealing results of all. In 2015 a paper was published in the, yes, very prestigious journal *Science* that carried the clickbait title "Evolutionary Resurrection of Flagellar Motility via Rewiring of the Nitrogen Regulation System."[15] Interestingly, the researchers had not intended to investigate the flagellum at all. Instead, they had wanted to study how a bacterium called *Pseudomonas fluorescens*—which normally colonizes plants and which in nature boasts a fine fully functioning flagellum—might cope with the problem of immobility. So they used laboratory techniques to delete *only* the master control gene that switches on synthesis of the molecular machine. All the dozens of other flagellar genes were left intact. After a few days the researchers were startled to see the kneecapped bacteria swimming once again. The science news website *The Scientist* hailed the result as a "giant evolutionary leap."[16]

The research results tell a different story, one that showcases the First Rule of Adaptive Evolution. It turns out that the system was switched back on by loss-of-FCT mutations. An alteration in one gene kept it turned on when it normally would be turned off (like a traffic light stuck on green), which then activated another gene too. The second gene was structurally similar to the master gene that controlled the flagellum, and its overactivation was enough to switch the system back on. Yet the bug was not a happy camper. As one of the authors noted: "The bacteria that became much better at swimming were much worse at nitrogen regulation," adding that "sometimes the advantage can be so great that it's worth paying that cost because otherwise you die."[17] I myself wouldn't call that an "evolutionary resurrection." On the other hand, a paper titled "Bacteria Endure Crippling Mutation to Stave Off Extinction" probably wouldn't be published in *Science*.

Almost all work on the evolution of flagella consists in comparing sequences, a method that—although it can support interesting conjectures about who descended from whom—says nothing about the mechanism of evolution. The very little work that has been done that's relevant to the mechanism strongly supports the arguments of this book.

Oops!

Besides the bacterial flagellum, another biochemical system discussed in *Darwin's Black Box* is the blood-clotting cascade, which will be the topic of this and the next sections. Although blood clotting seems simple on the surface—a small cut bleeds for a while, and then the bleeding slows down and eventually stops altogether—

biochemical investigations starting in the 1950s showed that it's remarkably complex, consisting of dozens of protein parts. In the cascade one protein activates the next, which activates the next, and so on. The complexity is needed not so much to coagulate blood as it is to control where and when coagulation happens. If a clot forms at the wrong time or place, it can cause a heart attack, stroke, or other health crisis.

I argued in the book that a major part of the blood-clotting cascade was irreducibly complex:[18] remove one of its necessary parts and the cascade breaks, either failing to clot at the right time or clotting at the wrong time. Thus it fits very poorly with Darwin's mechanism of evolution. In fact, at the time *Darwin's Black Box* was released, no science publication had yet shown how the blood-clotting system might have been produced by random mutation and natural selection.

That was my claim anyway. However, a man named Russell Doolittle (a distant relative of W. Ford Doolittle) disagreed, and in 1997 wrote an essay on the topic. That did not bode well for me, because Doolittle is a distinguished scientist—a (now retired) professor of biochemistry at the University of California–San Diego and member of the National Academy of Sciences who at that time had already been working on the blood-clotting cascade for forty years.

In his essay Doolittle made mostly standard evolutionary arguments (like those recounted in Chapters 3–5) with which I was very familiar and that caused me no concern. But one argument rattled me. Doolittle claimed that it had recently (that is, soon after my book was published, in a paper I had not yet read) been shown *experimentally* by other investigators that clotting was not irreducibly complex; specifically, that parts of the clotting cascade could be removed from mice with no ill effects. He wrote:

> Recently the gene for plaminogen [*sic*—plasminogen is a protein that helps remove blood clots after a wound has healed]

> was knocked out of mice, and, predictably, those mice had thrombotic complications because fibrin clots could not be cleared away. Not long after that, the same workers knocked out the gene for fibrinogen [fibrinogen supplies the protein building material for the meshwork clot structure] in another line of mice. Again, predictably, these mice were ailing, although in this case hemorrhage was the problem. And what do you think happened when these two lines of mice were crossed? For all practical purposes, the mice lacking both genes were normal! Contrary to claims about irreducible complexity, the entire ensemble of proteins is not needed. Music and harmony can arise from a smaller orchestra.[19]

In other words, the argument was that if one protein, plasminogen, is removed, mice have one set of problems. If a different protein, fibrinogen, is removed, they have a different set of problems. But if both proteins are removed, the mice are *normal*. So maybe the clotting cascade didn't have to arise all at once. Maybe it could have been built two proteins at a time—or something. Actually, Doolittle didn't address the problem of how exactly the clotting cascade could arise even if the experiments had shown what he thought they did. We needn't spend any time wondering about it, though, because it turns out that Doolittle had misread the paper. Mice missing both proteins are not normal at all, but very sick.

According to the investigators whose paper Doolittle cited, "Mice deficient in plasminogen and fibrinogen are phenotypically indistinguishable from fibrinogen-deficient mice."[20] Translated into English, that means mice missing both proteins have all the problems of mice that are missing only fibrinogen: their blood doesn't clot; they hemorrhage; females die during pregnancy. Promising evolutionary intermediates they are not.

Table A.1 shows a list of symptoms for the three lines of mice. Mice missing plasminogen have one set, mice missing fibrinogen

Table A.1. Symptoms of Mice with Gene Knockouts

Lacking plasminogen	Lacking fibrinogen	Lacking both
Thrombosis	No clotting	No clotting
Ulcers	Hemorrhage	Hemorrhage
High mortality	Death in pregnancy	Death in pregnancy

another. Mice missing both proteins are "rescued" (as the paper title put it) from the symptoms of plasminogen deficiency, but only to suffer the problems of fibrinogen deficiency. The reason is easy to understand. Fibrinogen is the precursor of the material of the clot itself, while plasminogen is the precursor of the protein that removes clots. So if a mouse can't make blood clots, it doesn't need plasminogen, because there are no clots that need to be cleared. Yet it still has all the problems that come from not being able to stop bleeding. The same group of researchers later separately investigated mice missing the blood-clotting proteins called prothrombin and tissue factor.[21] In both of those cases too the blood-clotting system broke down—which is exactly what you'd expect if the system were indeed irreducibly complex, as I had argued.

The point of this discussion is not that Professor Doolittle misread a paper. Anyone can do that. Scientific papers are not known for their clarity of prose; it took me a number of reads to puzzle out the paper too. Rather, there are two overriding lessons. The first lesson is that *experimental* work is of the utmost importance in determining just what Darwinian evolution can or can't do. We know for sure that those phantom "evolutionary intermediates" don't work only because researchers did the experiments that proved they didn't work. If, instead of citing a real experiment, Doolittle had only proposed some nebulous, hypothetical scenario for the origin of clotting, as other scientists have been wont to do, then critiques by skeptics like me would very likely have been disregarded by Darwinists. That's much harder to do with concrete lab results.

Russell Doolittle is one of the very top researchers in the area of blood clotting. Yet, as his essay clearly shows, he himself did not know how the clotting cascade could have arisen by Darwinian processes. Nor did he know of any papers in which an explanation had been given. If he had, he simply would have cited them. Instead, he cited a paper about dying mice. So the second important lesson from the affair is this: if Russell Doolittle himself cannot account for a system of the complexity of the clotting cascade by Darwinian processes, *nobody* can—nobody in the whole world.

Comparing Blood-Clotting Sequences

The hemorrhaging mice incident happened more than twenty years ago. Given the breakneck pace of science, has any progress been made since then in understanding how Darwin's mechanism might build a system such as the blood-clotting cascade? No, none at all. The literature is as barren of answers now as it was then. However, Russell Doolittle has continued to publish fascinating work on the cascade, which we'll examine here briefly.

Professor Doolittle has had a long, distinguished career working on many aspects of protein structure. He continued to work on blood clotting after the turn of the millennium, most especially by examining the burgeoning sequence data for the genomes of creatures on the evolutionary line from invertebrates to vertebrates. It turns out that humans and other mammals have nearly identical clotting cascades, with dozens of protein factors. On the other hand, animals without notochords (the spinal cords of vertebrates or a similar structure in certain invertebrates) have no clotting cascades. So somewhere along the phylogenetic line the clotting system must have arisen.

In the early 2000s genome data for the puffer fish became available. Puffer fish have jaws, which is a characteristic of the largest group of fish. On the other hand, lampreys—weird creatures with sucker mouths that attach themselves to prey and then rasp their tissue—belong to the group of jawless fish. Lampreys are thought to be descended from the earliest existing vertebrates on the evolutionary line leading to mammals. Their sequence data became available next. Like us, sea squirts are chordates, that is, they have notochords; unlike us, they don't have bones. Sea squirts are classified as tunicates—the closest relatives to vertebrates. Like the genome-sequence data of puffer fish and lampreys, sea squirt genome-sequence data also began to be available in the first decade of the 2000s. Doolittle and colleagues mined all that public information for whatever they could learn.[22]

In 2013 Russell Doolittle published a book, *The Evolution of Vertebrate Blood Clotting*, which summarized his findings and other work. In brief, he found that the simplest creatures he studied, tunicates, have some proteins whose sequences show a family resemblance to some clotting proteins of advanced vertebrates. Jawless fish have most of the proteins that mammals have. Jawed fish have all but a few. Other sequencing results show one or two blood-clotting factors missing from genomes of reptiles or birds. It's all fascinating work that required great dedication and erudition, and it is a good start at documenting what variations of the clotting cascade exist in nature. But virtually no experimental work was done, even on how the systems themselves work in the respective species, let alone on how they might change under selective pressure.

As I've said many times, although sequence similarities are good evidence for common descent, they cannot show whether random mutation and natural selection could build even the simplest system or, given that simple system, whether it could be expanded or improved by Darwin's mechanism. In 1992 biologist Torben Halkier aptly noted in a book on blood clotting:

> A system of this kind cannot just be allowed to free-wheel. The success of the coagulation process is due to the finely tuned modulation and regulation of all of the partial proteolytic digestions that occur. Too little or too much activity would be equally damaging for the organism. Regulation is a central issue in blood coagulation.[23]

Yet if *regulation* is a central issue in blood clotting, then any mutation to an already optimized system can be expected to disturb the balance. In the vast majority of circumstances that should be strongly opposed by natural selection. In order to even begin to understand how Darwinian processes might build a clotting cascade or even just significantly modify a preexisting one, huge roadblocks need to be addressed, such as how to maintain fine control on the fly while randomly changing a system. I wish luck to anyone with that.

What's more, as we've seen throughout this book, random mutation easily breaks or degrades genes. Since the blood-clotting cascade is a finely balanced system—a seesaw of opposing protein functions that either promote or inhibit clotting (Fig. A.2)—altering the balance by degrading one factor should be as effective in the short term as by strengthening another (like taking a bit of weight off one side of the seesaw instead of adding a bit to the other). And since degrading proteins is much faster and easier, that should almost always win out. As discussed in Chapter 9, the average time needed to evolve a mini–irreducibly complex feature such as a disulfide bond (a number of which occur in clotting proteins) is a million times that needed to degrade a gene.

As for Professor Doolittle, so too for the great majority of evolutionary biologists. All of these fundamental problems seem truly to be invisible to them. Evidence of common descent is routinely confused for evidence of Darwin's mechanism. Despite twenty years of efforts to refute the argument of *Darwin's Black Box*, not only has the

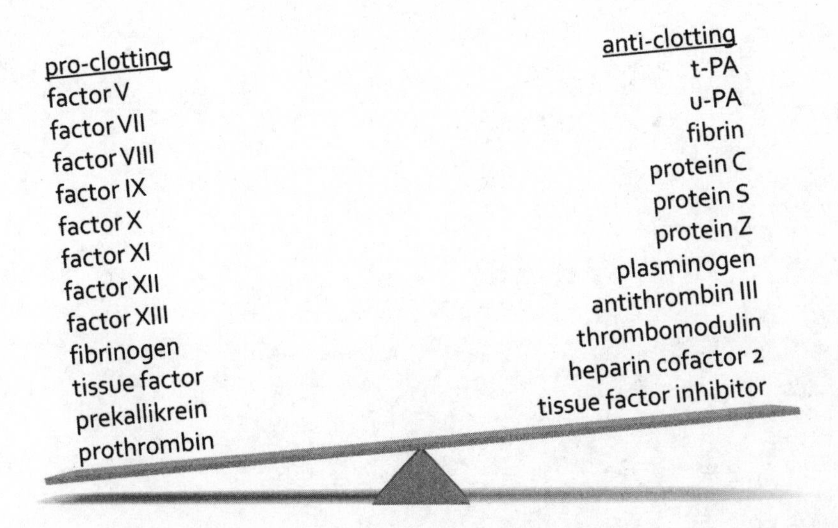

Figure A.2. The blood-clotting cascade seesaw, alternating between promoting and inhibiting coagulation. To change the balance, degrading one side would be very much quicker than strengthening the other side.

Darwinian evolution of the blood-clotting cascade not been solved; it hasn't even been addressed.

The utter sterility of Darwinian theory in accounting for complex functional systems (as shown by these and many other examples) should push evolutionary theorists to consider that they just may have been barking up the wrong tree.

ACKNOWLEDGMENTS

For helpful comments on draft chapters of the manuscript I gratefully acknowledge Douglas Axe, Richard Buggs, Enézio E. de Almeida Filho, Michael Egnor, Michael Flannery, Ann Gauger, Cornelius Hunter, Tony Jelsma, Casey Luskin, Jonathan McLatchie, Scott Minnich, Chase Nelson, Paul Nelson, Robert Sheldon, Richard Sternberg, and Jonathan Wells. Of course, their kind assistance does not imply that they necessarily endorse the book's argument, and any residual errors are entirely my own. I am also indebted to several reviewers who did not wish to be mentioned by name. Many thanks too to my editor, Katy Hamilton, at HarperOne as well as to my literary agent, Glen Hartley, of Writers' Representatives, for helping to put the manuscript into readable prose.

My sincere thanks goes to the folks at Discovery Institute for their terrific support over the decades, especially Bruce Chapman, Steve Meyer, John West, Casey Luskin, Rob Crowther, David Klinghoffer, and Janine Solfelt née Dixon. Thanks as well to my long-suffering colleagues in the Department of Biological Sciences at Lehigh University. Although publicly disagreeing with me,[1] they have also consistently upheld my right to express my controversial views and have sometimes had to endure astonishingly rude communications and other unpleasantries because of their principled stance.

Finally, I thank my impossibly wonderful wife, Celeste, without whom this book and very much else would not have been possible. Sweetie, we were designed for each other!

NOTES

Introduction

1. My discussion of ancient design arguments is based largely on A. S. Pease, "Caeli Enarrant," *Harvard Theological Review* 34 (1941): 163–200.

2. M. J. Schiefsky, "Galen's Teleology and Functional Explanation," *Oxford Studies in Ancient Philosophy* 33 (2007): 369–400.

3. W. Talbott, "Bayesian Epistemology," in E. N. Zalta, ed., *The Stanford Encyclopedia of Philosophy*, Winter 2016 ed.

4. A. R. Wallace, *The World of Life: A Manifestation of Creative Power, Directive Mind and Ultimate Purpose* (New York: Moffat, Yard, 1911).

5. L. J. Henderson, *The Fitness of the Environment* (New York: Macmillan, 1913).

6. For example, see M. J. Denton, *Nature's Destiny: How the Laws of Biology Reveal Purpose in the Universe* (New York: Free Press, 1998); G. F. Lewis and L. A. Barnes, *A Fortunate Universe: Life in a Finely Tuned Cosmos* (Cambridge: Cambridge Univ. Press, 2016).

7. T. Sommers and A. Rosenberg, "Darwin's Nihilistic Idea: Evolution and the Meaninglessness of Life," *Biology & Philosophy* 18 (2003): 653–68.

8. Sommers and Rosenberg, "Darwin's Nihilistic Idea."

9. E. Wasman and H. Muckermann, "Evolution," in *The Catholic Encyclopedia* (New York: Encyclopedia Press, 1909), 5: 654–70.

10. There have of course been many criticisms of intelligent design over the decades. Responses by myself and others can be found at the website of the Discovery Institute, www.discovery.org.

Chapter 1: The Pretense of Knowledge

1. S. Liu et al., "Population Genomics Reveal Recent Speciation and Rapid Evolutionary Adaptation in Polar Bears," *Cell* 157 (2014): 785–94.

2. Liu et al., "Population Genomics Reveal," Table S7.

3. By "biochemistry" I mean to include all disciplines that study life at the molecular level, such as molecular biology, biophysics, genetics, and others.

4. F. A. von Hayek, "The Pretence of Knowledge," December 11, 1974, https://www.nobelprize.org/nobel_prizes/economic-sciences/laureates/1974/hayek-lecture.html.

5. J. A. Coyne, "Of Vice and Men: The Fairy Tales of Evolutionary Psychology," a review of *A Natural History of Rape: Biological Bases of Sexual Coercion* by Randy Thornhill and Craig Palmer, *New Republic*, April 3, 2000.

6. C. Zimmer, "The Neurobiology of the Self," *Scientific American*, November 2005.

7. O. Judson, "The Hemiparasite Season," *New York Times*, December 25, 2014.

8. A. E. Johnson, "The Co-translational Folding and Interactions of Nascent Protein Chains: A New Approach Using Fluorescence Resonance Energy Transfer," *FEBS Letters* 579 (2005): 916–20.

9. D. Voet and J. G. Voet, *Biochemistry*, 4th ed. (New York: Wiley, 2011), p. 281.

10. K. Laland et al., "Does Evolutionary Theory Need a Rethink?" *Nature* 514 (2014): 161–64.

11. A. Wagner, *Arrival of the Fittest: Solving Evolution's Greatest Puzzle* (New York : Current, 2014); J. A. Shapiro, *Evolution: A View from the 21st Century* (Upper Saddle River, NJ: FT Press Science, 2011); E. V. Koonin, *The Logic of Chance: The Nature and Origin of Biological Evolution* (Upper Saddle River, NJ: FT Press Science, 2011); S. B. Gissis and E. Jablonka, eds., *Transformations of Lamarckism: From Subtle Fluids to Molecular Biology* (Cambridge, MA: MIT Press, 2011); M. Pigliucci and G. Müller, eds., *Evolution, the Extended Synthesis* (Cambridge, MA: MIT Press, 2010); P. J. Beurton, R. Falk, and H.-J. Rheinberger, eds., *The Concept of the Gene in Development and Evolution: Historical and Epistemological Perspectives* (Cambridge: Cambridge Univ. Press, 2008); M. Lynch, *The Origins of Genome Architecture* (Sunderland, MA: Sinauer, 2007); D. Noble, *The Music of Life* (Oxford: Oxford Univ. Press, 2006); R. A. Watson, *Compositional Evolution: The Impact of Sex, Symbiosis, and Modularity on the Gradualist Framework of Evolution* (Cambridge, MA: MIT Press, 2006); M. Kirschner and J. Gerhart, *The Plausibility of Life: Great Leaps of Evolution* (New Haven, CT: Yale Univ. Press, 2005); S. B. Carroll, *Endless Forms Most Beautiful: The New Science of Evo Devo and the Making of the Animal Kingdom* (New York: Norton, 2005); E. Jablonka and M. J. Lamb, *Evolution in Four Dimensions: Genetic, Epigenetic, Behavioral, and Symbolic Variation in the History of Life* (Cambridge, MA: MIT Press, 2005); M. J. West-Eberhard, *Developmental Plasticity and Evolution* (Oxford: Oxford Univ. Press, 2003); L. Margulis and D. Sagan, *Acquiring Genomes: A Theory of the Origins of Species* (New York: Basic Books, 2002).

12. Advertisement copy for J. A. Shapiro, *Evolution: A View from the 21st Century* at Amazon.com, https://www.amazon.com/Evolution-View-21st-Century-paperback/dp/0133435539.

13. Advertisement copy for E. V. Koonin, *The Logic of Chance: The Nature and Origin of Biological Evolution* at Amazon.com, https://www.amazon.com/Logic-Chance-Biological-Evolution-Science-ebook/dp/B0058I5U2C/ref=sr_1_1?s=books&ie=UTF8&qid=1532300761&sr=1-1&keywords=E.+V.+Koonin%2C+The+Logic+of+Chance%3A+The+Nature+and+Origin+of+Biological+Evolution.

14. Wagner, *Arrival of the Fittest*, p. 8.

15. P. Whoriskey, "The U.S. Government Is Poised to Withdraw Longstanding Warnings About Cholesterol," *Washington Post*, February 10, 2015.

16. G. Taubes, "The Soft Science of Dietary Fat," *Science* 291 (2001): 2536–45.

17. M. Wenner Moyer, "It's Time to End the War on Salt: The Zealous Drive by Politicians to Limit Our Salt Intake Has Little Basis in Science," *Scientific American*, July 8, 2011.

18. B. Vastag, "At USDA, a Plate Usurps the Food Pyramid," *Washington Post*, June 2, 2011.

19. R. A. Ferdman, "The Meat Industry's Worst Nightmare Could Soon Become a Reality," *Washington Post*, January 7, 2015.

20. *Ceteris paribus*, as philosophers say, "Everything else being equal." In other words, nothing substantive must change. For example, if the long jumper in the example strapped a rocket to his back or traveled to the moon for the second try, he might well clear 50 feet or more.

21. J. Gleick, *Chaos: Making a New Science* (London: Cardinal, 1987), p. 17.

22. The laws also typically have restricted ranges of application. For example, Newtonian gravity eventually yields to Einstein's theory of relativity.

23. H. F. Bunn, "The Triumph of Good over Evil: Protection by the Sickle Gene Against Malaria," *Blood* 121 (2013): 20–25.

24. A disturbing 2005 study claims that in fact the majority of published biomedical research findings are wrong due to bias and insufficient statistical rigor (J. P. Ioannidis, "Why Most Published Research Findings Are False," *PLoS Medicine* 2 [2005]: e124).

25. "Spandrel," www.thefreedictionary.com/spandrel.

26. M. Mitchell, *Complexity: A Guided Tour* (Oxford: Oxford Univ. Press, 2009), p. xii.

27. C. Darwin, *On the Origin of Species*, 1st ed. (1859; repr., New York: Bantam, 1999), p. 2.

Chapter 2: Fathomless Elegance

1. Quoted in C. Singer, *A History of Biology* (London: Abelard-Schuman, 1959), p. 4.

2. About the octopus, Aristotle wrote: "The octopus breeds in spring, lying hid for about two months. The female, after laying her eggs, broods over them. She thus gets out of condition since she does not go in quest of food during this time. The eggs are discharged into a hole and are so numerous that they would fill a vessel much larger than the animal's body. After about fifty days the eggs burst. The little creatures creep out, and are like little spiders, in great numbers. The characteristic form of their limbs are not yet visible in detail, but their general outline is clear. They are so small and helpless that the greater number perish. They have been so extremely minute as to be completely without organization, but nevertheless when touched they move" (quoted in Singer, *History of Biology*, p. 27).

3. D. Wootton, *Bad Medicine: Doctors Doing Harm Since Hippocrates* (Oxford: Oxford Univ. Press, 2006), pp. 135–38.

4. G. K. Chesterton, *Orthodoxy* (New York: John Lane, 1909), p. 51.

5. E. B. White and G. Williams, *Charlotte's Web* (New York: Harper & Row, 1952), p. 56.

6. M. Burrows and G. Sutton, "Interacting Gears Synchronize Propulsive Leg Movements in a Jumping Insect," *Science* 341 (2013): 1254–56.

7. E. Yong, "This Insect Has Gears in Its Legs," *National Geographic*, September 12, 2013, https://www.nationalgeographic.com/science/phenomena/2013/09/12/this-insect-has-gears-in-its-legs/.

8. C. Darwin, *On the Origin of Species*, 1st ed. (1859; repr., New York: Bantam, 1999), pp. 155–56.

9. Darwin, *Origin of Species*, pp. 157–58.

10. For example, see K. R. Miller, "Life's Grand Design," *Technology Review* 97 (1994): 24–32.

11. K. Franze et al., "Müller Cells Are Living Optical Fibers in the Vertebrate Retina," *Proceedings of the National Academy of Sciences USA* 104 (2007): 8287–92.

12. A. M. Labin et al., "Müller Cells Separate Between Wavelengths to Improve Day Vision with Minimal Effect upon Night Vision," *Nature Communications* 5 (2014): 4319.

13. J. Hewitt, "Fiber Optic Light Pipes in the Retina Do Much More Than Simple Image Transfer," *Phys.org*, July 21, 2014, https://phys.org/news/2014-07-fiber-optic-pipes-retina-simple.html.

14. R. Blakemore, "Magnetotactic Bacteria," *Science* 190 (1975): 377–79.

15. A. Komeili, "Molecular Mechanisms of Compartmentalization and Biomineralization in Magnetotactic Bacteria," *FEMS Microbiology Reviews* 36 (2012): 232–55.

16. D. Murat et al., "Comprehensive Genetic Dissection of the Magnetosome Gene Island Reveals the Step-wise Assembly of a Prokaryotic Organelle," *Proceedings of the National Academy of Sciences USA* 107 (2010): 5593–98.

17. J. S. Mattick, "Type IV Pili and Twitching Motility," *Annual Review of Microbiology* 56 (2002): 289–314.

18. B. Nan et al., "Bacteria That Glide with Helical Tracks," *Current Biology* 24 (2014): R169–73.

19. B. Nan et al., "Bacteria That Glide with Helical Tracks," R169–73.

20. The other two rotary motors belong to the bacterial flagellum and a protein complex called ATP synthase.

21. A. Shrivastava, P. P. Lele, and H. C. Berg, "A Rotary Motor Drives *Flavobacterium* Gliding," *Current Biology* 25 (2015): 338–41.

22. M. Miyata, "Unique Centipede Mechanism of Mycoplasma Gliding," *Annual Review of Microbiology* 64 (2010): 519–37.

23. S. V. Albers and K. F. Jarrell, "The Archaellum: How Archaea Swim," *Frontiers in Microbiology* 6 (2015): 1–12.

24. B. Alberts et al., *Molecular Biology of the Cell*, 6th ed. (New York: Garland Science, 2015), chap. 16.

25. N. W. Charon et al., "The Unique Paradigm of Spirochete Motility and Chemotaxis," *Annual Review of Microbiology* 66 (2012): 349–70.

26. J. Ruan et al., "Architecture of a Flagellar Apparatus in the Fast-Swimming Magnetotactic Bacterium MO-1," *Proceedings of the National Academy of Sciences USA* 109 (2012): 20643–48.

27. Other complications include regulation of gene expression by RNA and epigenetics.

Chapter 3: Synthesizing Evolution

1. J. Hutton, *Theory of the Earth* (London: Cadell, Junior, and Davies, 1795), p. 80.

2. C. Darwin, *On the Origin of Species*, 1st ed. (1859; repr., New York: Bantam, 1999), p. 400.

3. R. Owen, *Lectures on the Comparative Anatomy and Physiology of the Invertebrate Animals* (London: Longman, Brown, Green, and Longmans, 1843), p. 379.

4. Darwin, *Origin of Species*, p. 166.

5. Darwin's argument requires not only that the shared structure of limbs not be the best for each kind of animal, but also that it somehow was left unchanged by evolution. Yet if variation and selection could produce creatures as different as a whale and a bat, why was bone structure so difficult to improve? See M. Denton, *Evolution: Still a Theory in Crisis* (Seattle: Discovery Institute Press, 2016).

6. Further work has shown that the earliest cellular stages differ considerably. See M. K. Richardson et al., "There Is No Highly Conserved Embryonic Stage in the Vertebrates: Implications for Current Theories of Evolution and Development," *Anatomy and Embryology* (Berlin) 196 (1997): 91–106.

7. Darwin, *Origin of Species*, p. 365.

8. For his work on sequencing proteins and DNA Sanger won two Nobel prizes in chemistry (in 1958 and 1980).

9. R. C. Lewontin, "The Units of Selection," *Annual Review of Ecology and Systematics* 1 (1970): 1–18.

10. E. Mayr, *What Evolution Is* (New York: Basic Books, 2001), p. 188.

11. T. S. Kemp, *The Origin of Higher Taxa* (London and Chicago: Oxford Univ. Press and Univ. of Chicago Press, 2016).

12. Mayr, *What Evolution Is*, p. 205.

13. P. Ball, "The Strange Inevitability of Evolution," *Nautilus*, January 8, 2015.

14. Mayr, *What Evolution Is*, p. 86.

15. In the provision of variation, "chance rules supreme" (Mayr, *What Evolution Is*, p. 119).

16. C. N. Johnson, *Darwin's Dice: The Idea of Chance in the Thought of Charles Darwin* (New York: Oxford Univ. Press, 2015).

17. "Evolution is understood to be the result of an unguided, unplanned process of random variation and natural selection" (letter to the *New York Times* by 35 Nobel Laureates, September 9, 2005).

18. C. Darwin, in F. Darwin, ed., *The Life and Letters of Charles Darwin*, vol. 1 (New York: Appleton, 1897), pp. 278–79.

19. C. Darwin, letter to Asa Gray, May 22, 1860, *Darwin Correspondence Project*, https://www.darwinproject. ac.uk/letter/DCP-LETT-2814.xml. Darwin did allow, however, that God may have made the general laws of the universe.

20. "Consideration of how God could have carried out his task of Creation raised even more serious difficulties. The manifold adaptations of structure, activity, behavior, and life cycle for each of the millions of species of organisms were far too specific to be explained by general laws. On the other hand, it seemed quite unworthy of the creator

to believe that he personally arranged every detail in the traits and life cycles of every individual down to the lowest organism" (Mayr, *What Evolution Is*, p. 148).

21. W. Paley, *Natural Theology* (New York: American Tract Society, 1802), pp. 9–10. Some have argued that Paley plagiarized an earlier mathematician, Bernard Nieuwentyt, who published a very similar argument in 1716 in Dutch; see B. C. Jantzen, *An Introduction to Design Arguments* (New York: Cambridge Univ. Press, 2014), pp. 168–69.

22. H. B. Kettlewell, "Darwin's Missing Evidence," *Scientific American* 200 (1959): 48–53.

23. H. F. Bunn, "The Triumph of Good over Evil: Protection by the Sickle Gene Against Malaria," *Blood* 121 (2013): 20–25.

24. Darwin, *Origin of Species*, p. 400.

25. Mayr, *What Evolution Is*, p. 205.

Chapter 4: Magic Numbers

1. A. Stolzfus, "Why We Don't Want Another 'Synthesis,'" *Biology Direct* 12 (2017): 23, https://doi.org/10.1186/s13062-017-0194-1.

2. E. V. Koonin, *The Logic of Chance: The Nature and Origin of Biological Evolution* (Upper Saddle River, NJ: Pearson Education, 2012), p. vii.

3. C. Darwin, *On the Origin of Species*, 1st ed. (1859; repr., New York: Bantam, 1999), p. 71.

4. D. W. Coltman, et al., "Undesirable Evolutionary Consequences of Trophy Hunting," *Nature* 426 (2003): 655–58.

5. Darwin did allude to neutral features at one point: "This preservation of favourable variations and the rejection of injurious variations, I call Natural Selection. Variations neither useful nor injurious would not be affected by natural selection, and would be left as a fluctuating element, as perhaps we see in the species called polymorphic" (*Origin of Species*, p. 69).

6. M. Kimura, *The Neutral Theory of Molecular Evolution* (Cambridge: Cambridge Univ. Press, 1983).

7. Darwin, *Origin of Species*, p. 71.

8. Reliably, but not invariably. Like neutral ones, the great majority of beneficial mutations can be lost by chance processes before they are established.

9. The high rate of mutation leads some workers to worry about the viability of humanity's future. See M. Lynch, "Rate, Molecular Spectrum, and Consequences of Human Mutation," *Proceedings of the National Academy of Sciences USA* 107 (2010): 961–68; A. S. Kondrashov, *Crumbling Genome* (Hoboken, NJ: Wiley-Blackwell, 2017).

10. The advantageousness of a mutation is often represented in textbooks by a number called the selection coefficient, which can range from –1 through 0 to +1. –1 means the mutation is lethal, 0 means it's exactly neutral, and +1 means the mutation is necessary for survival (those without it perish). Mutations can have fractional values too. A selection coefficient of +0.1 means that the lucky mutant is expected to have one-tenth more heirs—eleven offspring for every ten that a nonmutant produces. A value of –0.1 means on average the mutant will produce only nine—one-tenth less. Those

differences may seem small, but over the generations the favorable mutation will take over the population. The unfavorable one will disappear.

As selection coefficients go, 0.1 is considered unusually strong. For mutations whose effects are weaker, the selection coefficient is assigned a number closer and closer to zero: 0.01, 0.001, 0.0001, and so on. Few mutations, perhaps none, are expected to have absolutely no effect on survival—that is, to be *exactly* equal to zero. But the exact value doesn't matter so much. Calculations anticipate that natural selection loses its powers of discrimination if the selection coefficient falls below a threshold value. These are called "nearly neutral" mutations.

If the effect of a mutation for good or ill is less than roughly the inverse of the population size (that is, 1 divided by the population size), it is effectively neutral. As an example, suppose a wimpy, barely harmful mutation would theoretically decrease the expected number of descendants of an organism over the generations by just one in a million (that is, its selection coefficient was –0.000001). If the population size were greater than a million, selection would disfavor it. Yet if there were fewer than a million organisms in the species, the mutation would behave as if it were neutral—selection wouldn't be able to see it.

11. M. Lynch, *The Origins of Genome Architecture* (Sunderland, MA: Sinauer, 2007).

12. The most portentous lucky events include multiple symbiotic associations of prokaryotic cells, such as the bacteria that became mitochondria.

13. M. Nei, *Mutation-Driven Evolution* (Oxford: Oxford Univ. Press, 2013), p. 155.

14. Nei, *Mutation-Driven Evolution*, p. 179.

15. R. Dawkins, "Inferior Design," *New York Times*, July 1, 2007.

16. Nei, *Mutation-Driven Evolution*, pp. 56–57.

17. Nei, *Mutation-Driven Evolution*, p. x.

18. Koonin, *Logic of Chance*, p. viii.

19. Koonin, *Logic of Chance*, p. 70.

20. Koonin, *Logic of Chance*, p. 102.

21. Koonin, *Logic of Chance*, p. 397.

22. Koonin, *Logic of Chance*, chap. 12.

23. Koonin, *Logic of Chance*, pp. 211, 242.

24. M. Mitchell, *Complexity: A Guided Tour* (Oxford: Oxford Univ. Press, 2009).

25. J. J. Tyson, "What Everyone Should Know About the Belousov-Zhabotinsky Reaction," in S. A. Levin, ed., *Frontiers in Mathematical Biology* (Berlin: Springer-Verlag, 1994), pp. 569–87.

26. A. Wagner, *Arrival of the Fittest: Solving Evolution's Greatest Puzzle* (New York: Current, 2014).

27. Nei, *Mutation-Driven Evolution*, p. 57.

Chapter 5: Overextended

1. K. N. Laland et al., "The Extended Evolutionary Synthesis: Its Structure, Assumptions and Predictions," *Proceedings of the Royal Society B* 282/1813 (2015), doi: 10.1098/rspb.2015.1019.

2. S. B. Carroll, *Endless Forms Most Beautiful: The New Science of Evo Devo and the Making of the Animal Kingdom* (New York: Norton, 2005).

3. M. Kirschner and J. Gerhart, *The Plausibility of Life: Resolving Darwin's Dilemma* (New Haven, CT: Yale Univ. Press, 2005).

4. E. Danchin et al., "Beyond DNA: Integrating Inclusive Inheritance into an Extended Theory of Evolution," *Nature Reviews Genetics* 12 (2011): 475–86.

5. P. Cubas, C. Vincent, and E. Coen, "An Epigenetic Mutation Responsible for Natural Variation in Floral Symmetry," *Nature* 401 (1999): 157–61.

6. A. Bird, "Perceptions of Epigenetics," *Nature* 447 (2007): 396–98.

7. E. I. Campos, J. M. Stafford, and D. Reinberg, "Epigenetic Inheritance: Histone Bookmarks Across Generations," *Trends in Cell Biology* 24 (2014): 664–74.

8. J. K. Arico et al., "Epigenetic Patterns Maintained in Early *Caenorhabditis elegans* Embryos Can Be Established by Gene Activity in the Parental Germ Cells," *PLoS Genetics* 7 (2011): e1001391.

9. J. P. Lim and A. Brunet, "Bridging the Transgenerational Gap with Epigenetic Memory," *Trends in Genetics* 29 (2013): 176–86.

10. K. Gapp et al., "Implication of Sperm RNAs in Transgenerational Inheritance of the Effects of Early Trauma in Mice," *Nature Neuroscience* 17 (2014): 667–69.

11. E. Jablonka and M. J. Lamb, "Transgenerational Epigenetic Inheritance," in M. Pigliucci and G. B. Müller, eds., *Evolution: The Extended Synthesis* (Cambridge, MA: MIT Press, 2010).

12. Jablonka and Lamb, "Transgenerational Epigenetic Inheritance," p. 164.

13. R. Dawkins, *The Blind Watchmaker* (New York: Norton, 1986), p. 9.

14. The first of these is sometimes called "ecosystem engineering" (D. H. Erwin, "Macroevolution of Ecosystem Engineering, Niche Construction and Diversity," *Trends in Ecology and Evolution* 23 [2008]: 304–10). Although the coiners of the phrase give no reason to think they mean intentional engineering activity by organisms, to an intelligent-design advocate like myself it is intriguing to think that large-scale alterations of the early earth, such as oxygenation of the atmosphere by photosynthetic bacteria, may represent purposeful terraforming of the planet in order to support subsequent advanced life-forms.

15. J. S. Turner, "Extended Phenotypes and Extended Organisms," *Biology and Philosophy* 19 (2004): 327–52.

16. Turner, "Extended Phenotypes."

17. Turner, "Extended Phenotypes."

18. Laland et al., "Extended Evolutionary Synthesis."

19. M. J. West-Eberhard, *Developmental Plasticity and Evolution* (New York: Oxford Univ. Press, 2003), p. 174.

20. M. Muschick et al., "Adaptive Phenotypic Plasticity in the Midas Cichlid Fish Pharyngeal Jaw and Its Relevance in Adaptive Radiation," *BMC Evolutionary Biology* 11 (2011): 116.

21. West-Eberhard, *Developmental Plasticity and Evolution*, p. 51.

22. West-Eberhard, *Developmental Plasticity and Evolution*, p. 29.

23. West-Eberhard, *Developmental Plasticity and Evolution*, p. 53.

24. West-Eberhard, *Developmental Plasticity and Evolution*, p. 51.

25. M. Pigliucci, "Phenotypic Plasticity," in M. Pigliucci and G. B. Müller, eds., *Evolution: The Extended Synthesis* (Cambridge, MA: MIT Press, 2010), p. 370.

26. J. A. Shapiro, *Evolution: A View from the 21st Century* (Upper Saddle River, NJ: FT Press Science, 2011), p. 134.

27. Shapiro, *Evolution*, p. 139.

28. Shapiro, *Evolution*, p. 134.

29. E. Szathmáry and J. Maynard Smith, "The Major Evolutionary Transitions," *Nature* 374 (1995): 227–32.

30. E. Szathmáry, "Toward Major Evolutionary Transitions Theory 2.0," *Proceedings of the National Academy of Sciences USA* 112 (2015): 10104–11.

31. J. Maynard Smith and E. Szathmáry, *The Major Transitions in Evolution* (Oxford, New York: Freeman Spektrum, 1995), p. xiii.

32. S. Mazur, "Steve Benner: Origins Soufflé, Texas-Style," *Huffington Post*, February 5, 2014, https://www.huffingtonpost.com/%20suzan-mazur/steve-benner-origins-souf_b_4374373.html.

33. K. Sterelny, "Evolvability Reconsidered," in B. Calcott and K. Sterelny, eds., *The Major Transitions in Evolution Revisited* (Cambridge, MA: MIT Press, 2011), pp. 83–100, 91–92.

34. Maynard Smith and Szathmáry, *Major Transitions in Evolution*, pp. 124–25.

35. Maynard Smith and Szathmáry, *Major Transitions in Evolution*, pp. 191–92.

36. Maynard Smith and Szathmáry, *Major Transitions in Evolution*, p. 192.

37. J. Maynard Smith, *Evolution and the Theory of Games* (Cambridge, New York: Cambridge Univ. Press, 1982).

38. P. A. Corning and E. Szathmáry, "'Synergistic Selection': A Darwinian Frame for the Evolution of Complexity," *Journal of Theoretical Biology* 371 (2015): 45–58.

Chapter 6: The Family Line

1. A. Stolzfus, "Why We Don't Want Another 'Synthesis,'" *Biology Direct* 12 (2017): 23, https://doi.org/10.1186/s13062-017-0194-1.

2. P. R. Grant and B. R. Grant, *How and Why Species Multiply: The Radiation of Darwin's Finches* (Princeton, NJ: Princeton Univ. Press, 2008), p. 58.

3. V. Vincek et al., "How Large Was the Founding Population of Darwin's Finches?" *Proceedings of the Royal Society B* 264 (1997): 111–18.

4. S. Lamichhaney et al., "Evolution of Darwin's Finches and Their Beaks Revealed by Genome Sequencing," *Nature* 518 (2015): 371–75.

5. S. Kleindorfer et al., "Species Collapse via Hybridization in Darwin's Tree Finches." *American Naturalist* 183 (2014): 325–41.

6. B. D. McKay and R. M. Zink, "Sisyphean Evolution in Darwin's Finches," *Biological Reviews* 90 (2014): 689–98.

7. Recently the Grants and coworkers have reported the apparent production of a new finch species by hybridization of two species from separate islands of the Galápagos; S. Lamichhaney, et al., "Rapid Hybrid Speciation in Darwin's Finches," *Science* 359 (2018): 224–28.

8. J. Weiner, *The Beak of the Finch: A Story of Evolution in Our Time* (New York: Knopf, 1994), p. 78.

9. A. Abzhanov et al., "BMP4 and Morphological Variation of Beaks in Darwin's Finches," *Science* 305 (2004): 1462–65.

10. A. Abzhanov et al., "The Calmodulin Pathway and Evolution of Elongated Beak Morphology in Darwin's Finches," *Nature* 442 (2006): 563–67.

11. Lamichhaney et al., "Evolution of Darwin's Finches."

12. E. Uz et al., "Disruption of ALX1 Causes Extreme Microphthalmia and Severe Facial Clefting: Expanding the Spectrum of Autosomal-Recessive ALX-Related Frontonasal Dysplasia," *American Journal of Human Genetics* 86 (2010): 789–96.

13. Lamichhaney et al., "Evolution of Darwin's Finches."

14. Formation of novel species from ancient, standing genetic variation has also been reported for finches from Australia and Papua New Guinea; K. F. Stryjewski and M. D. Sorenson, "Mosaic Genome Evolution in a Recent and Rapid Avian Radiation," *Nature Ecology & Evolution* 1 (2017): 1912–22.

15. IOC World Bird List, www.worldbirdnames.org.

16. D. H. Erwin et al., "The Cambrian Conundrum: Early Divergence and Later Ecological Success in the Early History of Animals," *Science* 334 (2011): 1091–97.

17. M. A. O'Leary et al., "The Placental Mammal Ancestor and the Post-K-Pg Radiation of Placentals," *Science* 339 (2013): 662–67.

18. J. G. M. Thewissen and E. M. Williams, "The Early Radiations of Cetacea (Mammalia): Evolutionary Pattern and Developmental Correlations," *Annual Review of Ecology and Systematics* 33 (2002): 73–90.

19. R. O. Prum et al., "A Comprehensive Phylogeny of Birds (Aves) Using Targeted Next-Generation DNA Sequencing," *Nature* 526 (2015): 569–73.

20. C. G. Sibley and B. L. Monroe, *Distribution and Taxonomy of Birds of the World* (New Haven, CT: Yale Univ. Press, 1990).

21. E. H. Davidson and D. H. Erwin, "Gene Regulatory Networks and the Evolution of Animal Body Plans," *Science* 311 (2006): 796–800.

22. This excludes the Caspian Sea, which is the largest body of water on earth enclosed by land. However, until about five million years ago, it was connected to the ocean.

23. G. W. Barlow, *The Cichlid Fishes: Nature's Grand Experiment in Evolution* (Cambridge, MA: Perseus, 2000).

24. Barlow, *Cichlid Fishes*, p. 24.

25. M. Friedman et al., "Molecular and Fossil Evidence Place the Origin of Cichlid Fishes Long After Gondwanan Rifting," *Proceedings of the Royal Society B* 280 (2013): 20131733.

26. D. Brawand et al., "The Genomic Substrate for Adaptive Radiation in African Cichlid Fish," *Nature* 513 (2014): 375–81, Extended Data Figure 2a.

27. H. Nagai et al., "Reverse Evolution in RH1 for Adaptation of Cichlids to Water Depth in Lake Tanganyika," *Molecular Biology and Evolution* 28 (2011): 1769–76.

28. S. Olson, *Evolution in Hawaii: A Supplement to Teaching About Evolution and the Nature of Science* (Washington, DC: National Academies Press, 2004).

29. J. K. Liebherr, "The *Mecyclothorax* Beetles (Coleoptera, Carabidae, Moriomorphini) of Haleakala, Maui: Keystone of a Hyperdiverse Hawaiian Radiation," *Zookeys* 544 (2015): 1–407.

30. T. J. Givnish et al., "Origin, Adaptive Radiation and Diversification of the Hawaiian Lobeliads (Asterales: Campanulaceae)," *Proceedings of the Royal Society B* 276 (2009): 407–16.

31. IOC World Bird List, www.worldbirdnames.org.

32. K. E. Nicholson et al., "It Is Time for a New Classification of Anoles (Squamata: Dactyloidae)," *Zootaxa* 3477 (2012): 1–108.

33. K. A. Jønsson et al., "Ecological and Evolutionary Determinants for the Adaptive Radiation of the Madagascan Vangas," *Proceedings of the National Academy of Sciences USA* 109 (2012): 6620–25.

34. A. D. Yoder, "Lemurs," *Current Biology* 17 (2007): R866–68.

35. Barlow, *The Cichlid Fishes*, p. 17.

Chapter 7: Poison-Pill Mutations

1. G. W. Barlow, *The Cichlid Fishes: Nature's Grand Experiment in Evolution* (Cambridge, MA: Perseus, 2000), p. 24.

2. S. F. Elena, V. S. Cooper, and R. E. Lenski, "Punctuated Evolution Caused by Selection of Rare Beneficial Mutations," *Science* 272 (1996): 1802–4.

3. R. E. Lenski et al., "Long-Term Experimental Evolution in *Escherichia coli*. I. Adaptation and Divergence During 2,000 Generations," *American Naturalist* 138 (1991): 1315–41.

4. F. Vasi, M. Travisano, and R. E. Lenski, "Long-Term Experimental Evolution in *Escherichia coli*. II. Changes in Life-History Traits During Adaptation to a Seasonal Environment," *American Naturalist* 144 (1994): 432–56.

5. Elena, Cooper, and Lenski, "Punctuated Evolution."

6. M. Travisano, F. Vasi, and R. E. Lenski, "Long-Term Experimental Evolution in *Escherichia coli*. III. Variation Among Replicate Populations in Correlated Responses to Novel Environments," *Evolution* 49 (1995): 189–200.

7. V. Souza, P. E. Turner, and R. E. Lenski, "Long-Term Experimental Evolution in *Escherichia coli*. V. Effects of Recombination with Immigrant Genotypes on the Rate of Bacterial Evolution," *Journal of Evolutionary Biology* 10 (1997): 743–69. This experiment was recently extended and the results analyzed in greater detail, with similar results; R. Maddamsetti and R. E. Lenski, "Analysis of Bacterial Genomes from an Evolution Experiment with Horizontal Gene Transfer Shows That Recombination Can Sometimes Overwhelm Selection," *PLoS Genetics* 14 (2018): e1007199.

8. P. D. Sniegowski, P. J. Gerrish, and R. E. Lenski, "Evolution of High Mutation Rates in Experimental Populations of *E. coli*," *Nature* 387 (1997): 703–5.

9. V. S. Cooper et al., "Mechanisms Causing Rapid and Parallel Losses of Ribose Catabolism in Evolving Populations of *Escherichia coli* B," *Journal of Bacteriology* 183 (2001): 2834–41.

10. Ten of 12 populations suffered deletions ranging from 0.9 percent to 3.5 percent of the ancestral genome size. The others duplicated some segments as well as deleting others. The authors write: "The overall tendency toward reduced genome size reflected the fact that large deletions were much more common than large duplications" (C. Raeside et al., "Large Chromosomal Rearrangements During a Long-Term Evolution Experiment with *Escherichia coli*," *MBio* 5 [2014]: e01377–14).

11. D. Schneider et al., "Long-Term Experimental Evolution in *Escherichia coli*. IX. Characterization of Insertion Sequence-Mediated Mutations and Rearrangements," *Genetics* 156 (2000): 477–88.

12. T. F. Cooper, D. E. Rozen, and R. E. Lenski, "Parallel Changes in Gene Expression After 20,000 Generations of Evolution in *Escherichia coli*," *Proceedings of the National Academy of Sciences USA* 100 (2003): 1072–77.

13. R. Dawkins, *The Greatest Show on Earth: The Evidence for Evolution* (New York: Free Press, 2009), pp. 124–25.

14. O. Tenaillon et al., "Tempo and Mode of Genome Evolution in a 50,000-Generation Experiment," *Nature* 536 (2016): 165–70.

15. More recent investigation by Lenski's lab suggests that mutations in a small minority (10 of 57) of selected *E. coli* genes may not completely break them but rather, as they put it, "fine-tune" them (probably by degrading their functions): "In some cases, adaptive mutations appear to modify protein functions, rather than merely knocking them out" (R. Maddamsetti et al., "Core Genes Evolve Rapidly in the Long-Term Evolution Experiment with *Escherichia coli*," *Genome Biology and Evolution* 9 [2017]: 1072–83).

16. M. J. Behe, "Experimental Evolution, Loss-of-Function Mutations, and 'The First Rule of Adaptive Evolution,'" *Quarterly Review of Biology* 85 (2010): 1–27.

17. I first realized that acronyms don't have to be slavishly restricted to just the initial letters of the words of a phrase by reading Calvin and Hobbes, who named their club G.R.O.S.S., for Get Rid Of Slimy girlS.

18. The terms "gain-of-function" and "loss-of-function" are common in evolutionary biology, but are ill-defined. I altered the terminology a bit to gain- and loss-of-FCT—that is, "Functional Coded elemenT"—to focus exclusively on whether coded elements are built or broken.

19. R. Carter and K. N. Mendis, "Evolutionary and Historical Aspects of the Burden of Malaria," *Clinical Microbiology Reviews* 15 (2002): 564–94.

20. S. Wielgoss et al., "Mutation Rate Dynamics in a Bacterial Population Reflect Tension Between Adaptation and Genetic Load," *Proceedings of the National Academy of Sciences USA* 110 (2013): 222–27.

21. Mutator strains come with short-term advantages and long-term disadvantages. See A. Giraud et al., "The Rise and Fall of Mutator Bacteria," *Current Opinion in Microbiology* 4 (2001): 582–85.

22. M. Kimura, "On the Evolutionary Adjustment of Spontaneous Mutation Rates," *Genetics Research* 9 (1967): 23–34.

23. M. J. Behe, "Getting There First: An Evolutionary Rate Advantage for Adaptive Loss-of-Function Mutations," in R. J. Marks et al., eds., *Biological Information: New Perspectives* (Singapore: World Scientific, 2013), pp. 450–73.

24. B. W. Wren, "The Yersiniae: A Model Genus to Study the Rapid Evolution of Bacterial Pathogens," *Nature Reviews Microbiology* 1 (2003): 55–64.

25. B. J. Hinnebusch et al., "Ecological Opportunity, Evolution, and the Emergence of Flea-Borne Plague," *Infection and Immunity* 84 (2016): 1932–40.

26. P. S. Chain et al., "Complete Genome Sequence of *Yersinia pestis* Strains Antiqua and Nepal516: Evidence of Gene Reduction in an Emerging Pathogen," *Journal of Bacteriology* 188 (2006): 4453–63.

27. Strain 91001 of *Y. pestis* is avirulent in humans but is lethal to mice; N. Thomson et al., "Brothers in Arms," *Nature Reviews Microbiology* 3 (2005): 100–101.

28. Z. D. Blount, C. Z. Borland, and R. E. Lenski, "Historical Contingency and the Evolution of a Key Innovation in an Experimental Population of *Escherichia coli*," *Proceedings of the National Academy of Sciences* USA 105 (2008): 7899–906.

29. The citrate mutation can be generated in weeks rather than decades by altering the bacterial growth regimen; D. J. Van Hofwegen, C. J. Hovde, and S. A. Minnich, "Rapid Evolution of Citrate Utilization by *Escherichia coli* by Direct Selection Requires citT and dctA," *Journal of Bacteriology* 198 (2016): 1022–34.

30. Z. D. Blount et al., "Genomic Analysis of a Key Innovation in an Experimental *Escherichia coli* Population," *Nature* 489 (2012): 513–18.

31. Blount et al., "Genomic Analysis," Supplementary Table 2.

32. Recent work by Lenski and colleagues strongly suggests that its fate will not be a happy one; A. Couce et al., "Mutator Genomes Decay, Despite Sustained Fitness Gains, in a Long-Term Experiment with Bacteria," *Proceedings of the National Academy of Sciences USA* 114 (2017): E9026–35, https://doi.org/10.1073/pnas.1705887114.

33. E. M. Quandt et al., "Fine-Tuning Citrate Synthase Flux Potentiates and Refines Metabolic Innovation in the Lenski Evolution Experiment," *eLife* (2015): 4.e09696.

34. Quandt et al., "Fine-Tuning Citrate Synthase Flux."

35. Quite recent research shows that the rapid advance in sequencing technology is now allowing comparative studies to track down adaptive gene loss in humans and other mammals. Sharma and colleagues write: "Our results suggest that gene loss is an evolutionary mechanism for adaptation that may be more widespread than previously anticipated. Hence, investigating gene losses has great potential to reveal the genomic basis underlying macroevolutionary changes." V. Sharma et al., "A Genomics Approach Reveals Insights into the Importance of Gene Losses for Mammalian Adaptations," *Nature Communications* 9 (2018): 1215–23. And Kronenberg and collaborators observe: "These findings are consistent with the 'less-is-more' hypothesis (61), which argues that the loss of functional elements underlies critical aspects of human evolution." Z. N. Kronenberg et al., "High-Resolution Comparative Analysis of Great Ape Genomes," *Science* 360 (2018): 1085–95.

36. A. K. Hottes et al., "Bacterial Adaptation Through Loss of Function," *PLoS Genetics* 9 (2013): e1003617.

37. Even though neither involved random mutations, the work of Hottes et al., "Bacterial Adaptation," is much more directly interpretable in evolutionary terms than an earlier experiment. The former tests the consequence of just one common type of mutation—breakage of a gene. The latter breaks genes and then guides the bacteria to the investigators' envisioned solution—overexpression of a separate *E. coli* gene (W. M. Patrick et al., "Multicopy Suppression Underpins Metabolic Evolvability," *Molecular Biology and Evolution* 24 [2007]: 2716–22).

38. A. Khare and S. Tavazoie, "Multifactorial Competition and Resistance in a Two-Species Bacterial System," *PLoS Genetics* 11 (2015): e1005715.

39. H. Stower, "Molecular Evolution: Adaptation by Loss of Function," *Nature Reviews Genetics* 14 (2013): 596.

40. D. Lutz, Washington Univ. in St. Louis, "How Repeatable Is Evolutionary History? 'Weakness' in Clover Genome Biases Species to Evolve Same Trait," *ScienceDaily*, June 23, 2014, https://www.sciencedaily.com/releases/2014/06/140623225009.htm.

41. K. M. Olsen, N. J. Kooyers, and L. L. Small, "Adaptive Gains Through Repeated Gene Loss: Parallel Evolution of Cyanogenesis Polymorphisms in the Genus *Trifolium* (Fabaceae)," *Philosophical Transactions of the Royal Society of London B Biological Sciences* 369 (2014): 20130347.

42. E. Yong, "One Gait-Keeper Gene Allows Horses to Move in Unusual Ways," *Discover*, August 29, 2012, blogs.discovermagazine.com/notrocketscience/2012/08/29/one-gait-keeper-gene-allows-horses-to-move-in-unusual-ways/#.Vs4eoPkrJhF.

43. L. S. Andersson et al., "Mutations in *DMRT3* Affect Locomotion in Horses and Spinal Circuit Function in Mice," *Nature* 488 (2012): 642–46.

44. J. Flannick et al., "Loss-of-Function Mutations in *SLC30A8* Protect Against Type 2 Diabetes," *Nature Genetics* 46 (2014): 357–63.

45. G. Kolata, "In Single Gene, a Path to Fight Heart Attacks," *New York Times*, June 18, 2014.

46. A more recent study of an Amish population shows another loss-of-function mutation that seems to prevent type 2 diabetes (S. S. Khan et al., "A Null Mutation in *SERPINE1* Protects Against Biological Aging in Humans," *Science Advances* 3 [2017], doi: 10.1126/sciadv.aao1617).

47. N. Wade, "Japanese Scientists Identify Ear Wax Gene," *New York Times*, January 29, 2006.

48. J. Ohashi, I. Naka, and N. Tsuchiya, "The Impact of Natural Selection on an ABCC11 SNP Determining Earwax Type," *Molecular Biology and Evolution* 28 (2011): 849–57.

49. R. Dawkins, "Inferior Design," *New York Times*, July 1, 2007.

50. J. J. Schoenebeck and E. A. Ostrander, "Insights into Morphology and Disease from the Dog Genome Project," *Annual Review of Cell and Developmental Biology* 30 (2014): 535–60.

51. D. S. Mosher et al., "A Mutation in the Myostatin Gene Increases Muscle Mass and Enhances Racing Performance in Heterozygote Dogs," *PLoS Genetics* 3 (2007): e79.

52. S. I. Candille et al., "A β-Defensin Mutation Causes Black Coat Color in Domestic Dogs," *Science* 318 (2007): 1418–23.

53. E. Cadieu et al., "Coat Variation in the Domestic Dog Is Governed by Variants in Three Genes," *Science* 326 (2009): 150–53.

54. M. Rimbault et al., "Derived Variants at Six Genes Explain Nearly Half of Size Reduction in Dog Breeds," *Genome Research* 23 (2013): 1985–95.

55. D. Bannasch et al., "Localization of Canine Brachycephaly Using an Across Breed Mapping Approach," *PLoS One* 5 (2010): e9632.

56. J. J. Schoenebeck and E. A. Ostrander, "The Genetics of Canine Skull Shape Variation," *Genetics* 193 (2013): 317–25.

57. I. Baranowska Körberg et al., "A Simple Repeat Polymorphism in the MITF-M Promoter Is a Key Regulator of White Spotting in Dogs," *PLoS One* 9 (2014): e104363.

58. K. Haworth et al., "Canine Homolog of the T-box Transcription Factor T; Failure of the Protein to Bind to Its DNA Target Leads to a Short-Tail Phenotype," *Mammalian Genome* 12 (2001): 212–18.

59. B. M. vonHoldt et al., "Structural Variants in Genes Associated with Human Williams-Beuren Syndrome Underlie Stereotypical Hypersociability in Domestic Dogs," *Science Advances* 3 (2017): e1700398.

60. C. D. Marsden et al., "Bottlenecks and Selective Sweeps During Domestication Have Increased Deleterious Genetic Variation in Dogs," *Proceedings of the National Academy of Sciences USA* 113 (2016): 152–57.

61. Cat lovers shouldn't feel vindicated—the only gene so far identified with the domestication of cats is loss-of-function of *Taqpep*, which disrupts tabby coat stripes, allowing other color patterns (C. Ottoni et al., "The Palaeogenetics of Cat Dispersal in the Ancient World," *Nature Ecology & Evolution* 1 [2017]: 0139). Similar disappointments await chicken lovers (L. G. Flink et al., "Establishing the Validity of Domestication Genes Using DNA from Ancient Chickens," *Proceedings of the National Academy of Sciences USA* 111 [2014]: 6184–89) and yeast lovers (S. W. Buskirk, R. E. Peace, and G. I. Lang, "Hitchhiking and Epistasis Give Rise to Cohort Dynamics in Adapting Populations," *Proceedings of the National Academy of Sciences USA* 114 [2017]: 8330–35).

62. S. Liu et al., "Population Genomics Reveal Recent Speciation and Rapid Evolutionary Adaptation in Polar Bears," *Cell* 157 (2014): 785–94.

63. V. J. Lynch et al., "Elephantid Genomes Reveal the Molecular Bases of Woolly Mammoth Adaptations to the Arctic," *Cell Reports* 12 (2015): 217–28.

64. Lynch et al., "Elephantid Genomes Reveal"; table available at https://usegalaxy.org/r/woolly-mammoth.

65. Another likely example is the flightless cormorant (A. Burga et al., "A Genetic Signature of the Evolution of Loss of Flight in the Galapágos Cormorant," *Science* 356 [2017], doi:10.1126/science. aal3345).

Chapter 8: Dollo's Timeless Law

1. *Online Etymology Dictionary*, s.v. "intelligence," https://www.etymonline.com/word/intelligence.

2. For example, D. J. Futuyma, "Natural Selection: How Evolution Works," *Actionbioscience*, December 2004, www.actionbioscience.org/evolution/futuyma.html.

3. J. T. Bridgham, S. M. Carroll, and J. W. Thornton, "Evolution of Hormone-Receptor Complexity by Molecular Exploitation," *Science* 312 (2006): 97–101.

4. S. J. Gould, "Dollo on Dollo's Law: Irreversibility and the Status of Evolutionary Laws," *Journal of the History of Biology* 3 (1970): 189–212.

5. J. T. Bridgham, E. A. Ortlund, and J. W. Thornton, "An Epistatic Ratchet Constrains the Direction of Glucocorticoid Receptor Evolution," *Nature* 461 (2009): 515–19.

6. It's interesting to note that in *The Edge of Evolution* I argued that if several amino-acid residues had to be changed before there was a beneficial effect, then that would likely be beyond the power of Darwinian evolution. The book was greeted with not so polite derision. Yet the Oregon work, even as it unintentionally confirms that conclusion, was published in a leading journal with laudatory news releases. Somewhere in there is a good project for a philosopher or at least a sociologist of science to ask why the same conclusion is met with howls of protest when presented by a Darwin skeptic, but with pats on the back when offered by a supporter of the theory.

7. Bridgham, Ortlund, and Thornton, "An Epistatic Ratchet."

8. Bridgham, Ortlund, and Thornton, "An Epistatic Ratchet."

9. Work by Thornton's lab on the DNA-recognition domain of estrogen- and steroid-binding receptors also shows that remarkably convoluted changes are required even for quite modest variations in their function (the switch of two bases in the recognition site, from AGGTCA to AGAACA); A. N. McKeown et al., "Evolution of DNA Specificity in a Transcription Factor Family Produced a New Gene Regulatory Module," *Cell* 159 [2014]: 58–68; D. W. Anderson, A. N. McKeown, and J. W. Thornton, "Intermolecular Epistasis Shaped the Function and Evolution of an Ancient Transcription Factor and Its DNA Binding Sites," *eLife* 4 [2015]: e07864; T. N. Starr, L. K. Picton, and J. W. Thornton, "Alternative Evolutionary Histories in the Sequence Space of an Ancient Protein," *Nature* 549 [2017]: 409–13).

10. J. Monod, *Chance and Necessity: An Essay on the Natural Philosophy of Modern Biology* (New York: Knopf, 1971).

11. More recent work shows that eleven neutral unselected serendipitous mutations are required for a more ancient version of a steroid receptor to give rise to more modern versions. Thornton calls those mutations "permissive." Perhaps a better word might be "anticipatory" (McKeown et al., "Evolution of DNA Specificity"; Starr, Picton, and Thornton, "Alternative Evolutionary Histories").

12. M. J. Behe, *Darwin's Black Box: The Biochemical Challenge to Evolution* (New York: Free Press, 1996), pp. 206–7.

13. H. Nagai et al., "Reverse Evolution in RH1 for Adaptation of Cichlids to Water Depth in Lake Tanganyika," *Molecular Biology and Evolution* 28 (2011): 1769–76.

14. S. C. Galen et al., "Contribution of a Mutational Hot Spot to Hemoglobin Adaptation in High-Altitude Andean House Wrens," *Proceedings of the National Academy of Sciences USA* 112 (2015): 13958–63.

15. M. J. Behe, *The Edge of Evolution: The Search for the Limits of Darwinism* (New York: Free Press, 2007), chap. 7.

16. Bridgham, Carroll, and Thornton, "Evolution of Hormone-Receptor Complexity."

17. R. A. Fisher, *The Genetical Theory of Natural Selection* (Oxford: Clarendon Press, 1930).

18. A. Force et al., "The Origin of Subfunctions and Modular Gene Regulation," *Genetics* 170 (2005): 433–46.

19. D. E. Nilsson and S. Pelger, "A Pessimistic Estimate of the Time Required for an Eye to Evolve," *Proceedings of the Royal Society B Biological Sciences* 256 (1994): 53–58.

20. R. E. Lenski et al., "The Evolutionary Origin of Complex Features," *Nature* 423 (2003): 139–44.

21. S. F. Elena and R. E. Lenski, "Evolution Experiments with Microorganisms: The Dynamics and Genetic Bases of Adaptation," *Nature Reviews Genetics* 4 (2003): 457–69.

22. G. Zhu, G. B. Golding, and A. M. Dean, "The Selective Cause of an Ancient Adaptation," *Science* 307 (2005): 1279–82.

23. P. R. Grant and B. R. Grant, "Unpredictable Evolution in a 30-Year Study of Darwin's Finches," *Science* 296 (2002): 707–11.

24. R. Dawkins, "Inferior Design," *New York Times*, July 1, 2007.

25. R. Dawkins, "Conversation: Who Is the Greatest Biologist of All Time?" https://www.edge.org/conversation/who-is-the-greatest-biologist-of-all-time.

26. Fisher, *Genetical Theory of Natural Selection*, pp. 95–96.

27. G. P. Wagner, "The Changing Face of Evolutionary Thinking," *Genome Biology and Evolution* 5 (2013): 2006–7.

28. R. MacIntyre, "Book Review: *Mutation-Driven Evolution*," *Journal of Heredity* 106 (2015): 420.

Chapter 9: Revenge of the Principle of Comparative Difficulty

1. C. Darwin, *On the Origin of Species*, 1st ed. (1859; repr., New York: Bantam, 1999), p. 157.

2. Darwin, *Origin of Species*, p. 156.

3. For example, see R. Dawkins, *Climbing Mount Improbable* (New York: Norton, 1996), chap. 5; E. Mayr, *What Evolution Is* (New York: Basic Books, 2001), pp. 204–7; D. E. Nilsson and S. Pelger, "A Pessimistic Estimate of the Time Required for an Eye to Evolve," *Proceedings of the Royal Society B Biological Sciences* 256 (1994): 53–58.

4. Darwin, *Origin of Species*, p. 162.

5. Darwin, *Origin of Species*, p. 158.

6. M. J. Behe, *Darwin's Black Box: The Biochemical Challenge to Evolution* (New York: Free Press, 1996), p. 39.

7. R. Pennock, *Tower of Babel: The Evidence Against the New Creationism* (Cambridge, MA: MIT Press, 1999).

8. R. H. Thornhill and D. W. Ussery, "A Classification of Possible Routes of Darwinian Evolution," *Journal of Theoretical Biology* 203 (2000): 111–16.

9. J. McDonald, "A Reducibly Complex Mousetrap," January 13, 2003, udel .edu/~mcdonald/oldmousetrap.html.

10. M. J. Behe, "Irreducible Complexity: Obstacle to Darwinian Evolution," in W. A. Dembski and M. Ruse, eds., *Debating Design: From Darwin to DNA* (Cambridge: Cambridge Univ. Press, 2004), pp. 352–70.

11. That was essentially Ernst Mayr's tack (discussed in Chapter 3) when he wrote that the existence of different kinds of eyes in nature "refute[s] the claim that the gradual evolution of a complex eye is unthinkable." Any skeptic who accepts the burden to prove to the satisfaction of folks with remarkably elastic imaginations that such a transformation is unthinkable has taken on an impossible task.

12. B. Alberts, "The Cell as a Collection of Protein Machines: Preparing the Next Generation of Molecular Biologists," *Cell* 92 (1998): 291–94.

13. J. M. Smith, "Natural Selection and the Concept of a Protein Space," *Nature* 225 (1970): 563–64; W. H. Li, *Molecular Evolution* (Sunderland, MA: Sinauer Associates, 1997), p. 427; H. A. Orr, "A Minimum on the Mean Number of Steps Taken in Adaptive Walks," *Journal of Theoretical Biology* 220 (2003): 241–47.

14. M. J. Behe and D. W. Snoke, "Simulating Evolution by Gene Duplication of Protein Features That Require Multiple Amino Acid Residues," *Protein Science* 13 (2004): 2651–64.

15. M. Kimura, "The Role of Compensatory Neutral Mutations in Molecular Evolution," *Journal of Genetics* 64 (1985): 7–19; T. Ohta, "Time for Spreading of Compensatory Mutations Under Gene Duplication," *Genetics* 123 (1989): 579–84.

16. M. Lynch, "Simple Evolutionary Pathways to Complex Proteins," *Protein Science* 14 (2005): 2217–25.

17. M. J. Behe and D. W. Snoke, "A Response to Michael Lynch," *Protein Science* 14 (2005): 2226–27.

18. That assumes a mutation rate of 10^{-8} per generation and a selection coefficient of 0.01.

19. Lynch, "Simple Evolutionary Pathways," Fig. 3, for a selection coefficient of 0.01 and target size of two residues, such as that used in Behe and Snoke, "Simulating Evolution by Gene Duplication."

20. Behe and Snoke, "Simulating Evolution by Gene Duplication," Fig. 6.

21. D. J. Whitehead et al., "The Look-Ahead Effect of Phenotypic Mutations," *Biology Direct* 3 (2008): 18.

22. E. V. Koonin, reviewer's comment in Whitehead et al., "The Look-Ahead Effect."

23. Although phenotypic mutations might have a small effect on which protein sequence is the overall best for an organism (see S. Bratulic, F. Gerber, and A. Wagner, "Mistranslation Drives the Evolution of Robustness in TEM-1 β-Lactamase," *Proceedings of the National Academy of Sciences USA* 112 [2015]: 12758–63), they aren't at all, to mix metaphors, the blind watchmaker's magic bullet that Koonin thought. See M. J. Behe, "The Old Enigma," *Uncommon Descent*, April 2009, https://web.archive.org/web /20090816035853/http://behe.uncommondescent.com/.

24. G. K. Chesterton, *Orthodoxy* (New York: John Lane, 1909), chap. 2.

25. Darwinian evolution seems to explain both political liberalism and conservatism (P. Singer, *A Darwinian Left: Politics, Evolution and Cooperation* [New Haven, CT: Yale Univ. Press, 2000]; L. Arnhart, *Darwinian Conservatism* [Exeter: Imprint Academic, 2005]).

26. G. E. White, *Justice Oliver Wendell Holmes: Law and the Inner Self* (New York: Oxford Univ. Press, 1993).

27. J. Carroll, *Literary Darwinism: Evolution, Human Nature, and Literature* (New York: Routledge, 2004).

28. N. L. Wallin, B. Merker, and S. Brown, *The Origins of Music* (Cambridge, MA: MIT Press, 2000).

29. D. Maestripieri, "The Evolutionary History of Love: What Love Is and Where It Comes From," *Psychology Today*, March 26, 2012, https://www.psychologytoday.com/us/blog/games-primates-play/201203/the-evolutionary-history-love.

30. L. Smolin, *The Life of the Cosmos* (New York: Oxford Univ. Press, 1997).

31. D. C. Dennett, *Darwin's Dangerous Idea: Evolution and the Meanings of Life* (New York: Simon & Schuster, 1995).

32. For a model for estimating how much more quickly degradative mutations would arrive and fix in a species compared to single constructive mutations, see M. J. Behe, "Getting There First: An Evolutionary Rate Advantage for Adaptive Loss-of-Function Mutations," in R. J. Marks et al., eds., *Biological Information: New Perspectives* (Singapore: World Scientific, 2013), pp. 450–73. The case for mIC features would be much worse.

33. Lynch, "Simple Evolutionary Pathways."

34. Darwin set the example, rationalizing why new life no longer arises from inorganic matter, as it presumably did in the past: "It is often said that all the conditions for the first production of a living organism are now present, which could ever have been present.— But if (& oh what a big if) we could conceive in some warm little pond with all sorts of ammonia & phosphoric salts,—light, heat, electricity &c present, that a protein compound was chemically formed, ready to undergo still more complex changes, at the present day such matter wd be instantly devoured, or absorbed, which would not have been the case before living creatures were formed" (C. Darwin, letter to J. D. Hooker, February 1, 1871, *Darwin Correspondence Project*, https://www.darwinproject.ac.uk/letter/DCP-LETT-7471.xml).

35. L. A. Hug et al., "A New View of the Tree of Life," *Nature Microbiology* 1 (2016): 16048.

Chapter 10: A Terrible Thing to Waste

1. C. Darwin, *On the Origin of Species*, 1st ed. (1859; repr., New York: Bantam, 1999), p. 71.

2. R. Kirk, "Zombies," in E. N. Zalta, ed., *The Stanford Encyclopedia of Philosophy* (Summer 2015 Edition), https://plato.stanford.edu/archives/sum2015/entries/zombies.

3. F. Crick, *The Astonishing Hypothesis: The Scientific Search for the Soul* (New York: Scribner; Maxwell Macmillan International, 1994), p. 3.

4. Imagine if at the outset, instead of projecting his idea onto his readers, Crick had declared: "The Astonishing Hypothesis is that 'I,' my joys and my sorrows, are in fact no more than the behavior of a vast assembly of nerve cells. I'm nothing but a pack of neurons." I do believe his readers would have lost confidence in him.

5. C. Darwin, in F. Darwin, ed., *The Life and Letters of Charles Darwin*, vol. 1 (New York: Appleton, 1897), pp. 278–79.

6. D. J. Futuyma, "Natural Selection: How Evolution Works," *Actionbioscience*, December 2004, www.actionbioscience.org/evolution/futuyma.html.

7. J. R. Searle, *Mind: A Brief Introduction* (Oxford: Oxford Univ. Press, 2004), chap. 8.

8. T. Reid, *Essays on the Intellectual Powers of Man* (New York: Garland, 1971), chap. 6.

9. Much more could be said here that is beyond the scope of this book. Just as one example, if we already know that other minds exist, we can deduce that such a mind is behind some spandrel of intelligence—the unintentional side effects of purposely arranged parts—such as, say, a vapor trail left in the sky by a jet.

10. It may seem surprising, but research shows that most persons with locked-in syndrome are quite happy (M.-A. Bruno et al., "A Survey on Self-Assessed Well-Being in a Cohort of Chronic Locked-In Syndrome Patients: Happy Majority, Miserable Minority," *BMJ Open* 1 [2011]: e000039, doi:10.1136/ bmjopen-2010–000039).

11. Other ways of recognizing the work of an intelligent mind turn out to depend on the purposeful arrangement of parts. For example, a colleague of mine, Stephen Meyer, of the Discovery Institute, has written that we perceive intelligence in information. But how do we know that something such as a string of letters is information? After all, monkeys banging on a typewriter will put out a string of letters, but no information. The key is perceiving that the letters have been arranged for a purpose, the purpose of spelling out and communicating a message.

12. R. Dawkins, *The Blind Watchmaker* (New York: Norton, 1986), p. 1.

13. Dawkins, *Blind Watchmaker*, p. 21.

14. Dawkins, *Blind Watchmaker*, p. 21.

15. "Science," *The Free Dictionary*, www.thefreedictionary.com/science.

16. C. Singer, *A History of Biology* (London: Abelard-Schuman, 1959), p. 114.

17. P. Guyer and R. Horstmann, "Idealism," in E. N. Zalta, ed., *The Stanford Encyclopedia of Philosophy* (Fall 2015 Edition), https://plato.stanford.edu/archives/fall2015/entries/idealism.

18. C. Darwin, letter to William Graham, July 3, 1881, *Darwin Correspondence Project*, https://www.darwinproject.ac.uk/letter/DCP-LETT-13230.xml.

19. P. Churchland, "Epistemology in the Age of Neuroscience," *Journal of Philosophy* 84 (1987): 544–53.

20. W. Ramsey, "Eliminative Materialism," in E. N. Zalta, ed., *The Stanford Encyclopedia of Philosophy* (Summer 2013 Edition), https://plato.stanford.edu/archives/sum2013/entries/materialism-eliminative.

21. Ramsey, "Eliminative Materialism."

22. N. Bostrom, *Anthropic Bias: Observation Selection Effects in Science and Philosophy* (New York: Routledge, 2002), p. 11.

23. Bostrom, *Anthropic Bias*, pp. 52–53, 55.

24. N. Bostrum, "Are We Living in a Computer Simulation?" *Philosophical Quarterly* 53 (2003): 243–55.

25. American Museum of Natural History, "2016 Isaac Asimov Memorial Debate: Is the Universe a Simulation?" https://www.amnh.org/explore/news-blogs/podcasts/2016-isaac-asimov-memorial-debate-is-the-universe-a-simulation.

26. C. Moskowitz, "Are We Living in a Computer Simulation?" *Scientific American*, April 7, 2016, https://www.scientificamerican.com/article/are-we-living-in-a-computer-simulation.

27. Moskowitz, "Are We Living in a Computer Simulation?"

28. For example, the late, distinguished scholar of science and religion Ian Barbour wrote: "Why was it in *Western civilization* alone, among all the cultures of the world, that science in its modern form developed? . . . The medieval legacy also included presuppositions about nature that were congenial to the scientific enterprise. First, the conviction of *the intelligibility of nature* contributed to the rational or theoretical component of science. The medieval scholastics, like the Greek philosophers, did have great confidence in human rationality. Moreover, they combined the Greek view of the orderliness and regularity of the universe with the biblical view of God as Lawgiver. Monotheism implies the universality of order and coherence. . . . Second, the doctrine of creation implies that the details of nature can be known only by observing them. For if the world is the product of God's free act, it did not have to be made as it was made, and we can understand it only by actual observation. The universe, in other words, is contingent on God's will, not a necessary consequence of first principles. This world is both orderly and contingent, for God is both rational and free" (emphasis in original; I. G. Barbour, *Religion and Science: Historical and Contemporary Issues* [San Francisco: HarperSanFrancisco, 1997], pp. 27–28).

29. C. Darrow, "Closing Argument," *The State of Illinois v. Nathan Leopold & Richard Loeb*, August 22, 1924, law2.umkc.edu/faculty/projects/ftrials/leoploeb/darrowclosing.html.

30. Searle, *Mind*, chap. 3.

31. A. Koestler, *The Ghost in the Machine* (London: Hutchinson, 1967).

32. T. Nagel, *Mind and Cosmos: Why the Materialist Neo-Darwinian Conception of Nature Is Almost Certainly False* (New York: Oxford Univ. Press, 2012), pp. 3, 16.

33. M. J. Behe, "The Pilgrim's Regress: A Review of Richard Dawkins's *The Ancestor's Tale*," *American Spectator*, April 2005, pp. 54–57.

34. R. Dawkins, *The Ancestor's Tale: A Pilgrimage to the Dawn of Evolution* (Boston: Houghton Mifflin, 2004), p. 6.

35. Charles Simonyi was one of the early developers of Microsoft computer applications.

Appendix

1. On October 4, 2007, the Council of Europe's Parliamentary Assembly approved a resolution urging its member governments to oppose the teaching of creationism as science. Like many such official statements, the resolution recklessly conflates intelligent design with creationism: "The 'intelligent design' idea, which is the latest,

more refined version of creationism, does not deny a certain degree of evolution. However, intelligent design, presented in a more subtle way, seeks to portray its approach as scientific, and therein lies the danger." Apparently, thinking subtly and making distinctions about evolution are strongly discouraged. The draft resolution was ultimately not approved (http://assembly.coe.int/nw/xml/XRef/Xref-XML2HTML-EN .asp?fileid=17592&lang=en).

2. M. J. Pallen, and N. J. Matzke, "From the *Origin of Species* to the Origin of Bacterial Flagella," *Nature Reviews Microbiology* 4 (2006): 784–90.

3. E. Mayr, *What Evolution Is* (New York: Basic Books, 2001), p. 148.

4. M. Beeby et al., "Diverse High-Torque Bacterial Flagellar Motors Assemble Wider Stator Rings Using a Conserved Protein Scaffold," *Proceedings of the National Academy of Sciences USA* 113 (2016): E1917–26.

5. Pallen and Matzke, "From the *Origin of Species*."

6. Mayr, *What Evolution Is*, p. 148.

7. Great examples of nontrivial variants of the "core" flagellum have recently been characterized by B. Chaban and colleagues (B. Chaban, I. Coleman, and M. Beeby, "Evolution of Higher Torque in Campylobacter-Type Bacterial Flagellar Motors," *Scientific Reports* 8 [2018]: 97). However, although the authors write of the "evolution" of the variants, no experiments were done to test whether Darwinian processes were competent to accomplish the required transitions. Rather, as with Pallen and Matzke ("From the *Origin of Species*"), that was simply assumed.

8. M. J. Pallen, and U. Gophna, "Bacterial Flagella and Type III Secretion: Case Studies in the Evolution of Complexity," *Genome Dynamics* 3 (2007): 30–47; L. A. Snyder et al., "Bacterial Flagellar Diversity and Evolution: Seek Simplicity and Distrust It?" *Trends in Microbiology* 17 (2009): 1–5.

9. J. Curtaro, "A Complex Tail, Simply Told," *ScienceNOW*, April 17, 2007, http://www.sciencemag.org/news/2007/04/complex-tail-simply-told.

10. In 2005 the ACLU sued the school board of the tiny town of Dover, Pennsylvania. The school board had mandated that a (surprisingly poorly written) statement concerning evolution be read to high-school students in biology class. The statement was frankly skeptical of Darwin's theory and informed the students that there was a book in the school library called *Of Pandas and People* that discussed intelligent design. The plaintiffs complained that was an intrusion of religion into the public schools.

After the federal trial the judge, a man named John Jones (whose background had been as a lawyer, unsuccessful candidate for Congress, and a politically appointed head of the Pennsylvania Liquor Control Board) ruled for the plaintiffs. His written opinion garnered much praise from all the right people, including the staff at *Time* magazine, who honored Jones with a place on its 2006 *Time* 100 list of the "100 men and women whose power, talent or moral example is transforming our world." Yet there is little reason to think he comprehended any of the expert testimony at the trial, for either the plaintiffs or the defendants (I was a witness for the defense). For example, here's an excerpt from the opinion that put Judge Jones on the cover of *Time* magazine:

> Indeed, the assertion that design of biological systems can be inferred from the "purposeful arrangement of parts" is based upon an analogy to human design. Because we are able to recognize design of artifacts and objects, according to Professor Behe, that same reasoning can be employed to determine biological design.

And here's an excerpt from a plaintiffs' lawyers' document, which had earlier been given to the judge:

> The assertion that design of biological systems can be inferred from the "purposeful arrangement of parts" is based on an analogy to human design. According to Professor Behe, because we are able to recognize design of artifacts and objects, that same reasoning can be employed to determine biological design.

Jones simply *copied the text he was handed*. Whenever the topic concerned the testimony of the expert witnesses—whether scientists, philosophers, or theologians, whether for the plaintiffs or for the defendants—the very same language from the lawyers' document was inserted into the opinion with copy-and-paste efficiency, sometimes very lightly copyedited. Tellingly, when the opinion shifted to mundane matters, like school-board meetings or local newspaper editorials, he used his own voice; he was apparently perfectly comfortable writing for himself on those topics.

Now, let's ask ourselves why lifting material from somebody else is a bad idea. Why are reporters and politicians disgraced if they're caught doing so? Perhaps more to the point, why are students at all levels taught it's very wrong to plagiarize the work of another person? One reason is that a teacher wants to see if a student understands a topic and is able to restate arguments in such a way that indicates comprehension. After all, anyone could copy from a book on a difficult topic such as, say, quantum mechanics or Aristotle's *Metaphysics*. But copying the text surely doesn't show that the person understands the material. Apparently the legal system exempts itself from the standards that the rest of us follow.

The dilemma, however, remains. If a judge simply copies a text on a complicated matter, there's no evidence to show he understands it. In such a case the losing side may justifiably suspect that it didn't really get its day in court—that the judge's choice of text to copy had little to do with the inherent logic of the arguments. What's more, in the particular case of the Dover trial there's every reason to think that John Jones was completely at sea. (Detailed critiques of the opinion can be found on the website of the Discovery Institute.) It is quite doubtful that the former liquor-store bureaucrat grasped the distinction between such pivotal technical topics as flagellin and prothrombin, gene duplication and point mutation, Thomas Aquinas and David Hume, or random mutation and common descent. Thus, while the ruling of religious motivation on the part of the school board may have been correct as a legal matter, as a comment on the academic content of intelligent design the opinion was vacuous. A courtroom is no place to discuss intellectual issues.

11. N. Matzke, "Flagellum Evolution Paper Exhibits Canine Qualities," *The Panda's Thumb*, April 16, 2007, https://pandasthumb.org/archives/2007/04/flagellum-evolu-1.html.

12. W. F. Doolittle and O. Zhaxybayeva, "Reducible Complexity—The Case for Bacterial Flagella," *Current Biology* 17 (2007): R510–12.

13. Here's an example that in my experience is all too typical. Alan Leshner, long-time head of the American Association for the Advancement of Science, wrote an editorial in 2002 decrying intelligent design and proudly noting that "the Board of Directors of the American Association for the Advancement of Science (AAAS), the largest general scientific society in the world, passed a resolution this month urging policy-makers to keep intelligent design theory out of U.S. science classrooms" (A. I. Leshner, "'Intelligent Design' Theory Threatens Science Classrooms," *Seattle Post-Intelligencer*, November 22, 2002).

John West, of the Discovery Institute, decided to ask a few questions: "I wrote to Leshner and other members of the board asking them what books or articles by scientists favoring intelligent design they had read before adopting their resolution. Leshner declined to identify any and replied instead that the issue had been analyzed by his group's policy staff. Another board member similarly declined to specify anything she had read by design proponents, while a third board member volunteered that she had perused unspecified sources on the Internet. In other words, it appears that board members voted to brand intelligent design as unscientific without actually reading for themselves the academic books and articles by scientists proposing the theory" (J. G. West, "Intelligent Design Could Offer Fresh Ideas on Evolution," *Seattle Post-Intelligencer*, December 6, 2002, https://www.discovery.org/a/1313).

14. Doolittle and Zhaxybayeva, "Reducible Complexity."

15. T. B. Taylor et al., "Evolutionary Resurrection of Flagellar Motility via Rewiring of the Nitrogen Regulation System," *Science* 347 (2015): 1014–17.

16. R. Williams, "Evolutionary Rewiring: Strong Selective Pressure Can Lead to Rapid and Reproducible Evolution in Bacteria," *The Scientist*, February 26, 2015, https://www.the-scientist.com/daily-news/evolutionary-rewiring-35878.

17. Williams, "Evolutionary Rewiring."

18. In *Darwin's Black Box* I restricted my argument to the part of the cascade after the fork between the intrinsic and extrinsic branches.

19. R. F. Doolittle, "A Delicate Balance," *Boston Review*, February–March 1997, http://bostonreview.net/archives/BR22.1/doolittle.html.

20. T. H. Bugge et al., "Loss of Fibrinogen Rescues Mice from the Pleiotropic Effects of Plasminogen Deficiency," *Cell* 87 (1996): 709–19.

21. T. H. Bugge et al., "Fatal Embryonic Bleeding Events in Mice Lacking Tissue Factor, the Cell-Associated Initiator of Blood Coagulation," *Proceedings of the National Academy of Sciences USA* 93 (1996): 6258–63; W. Y. Sun et al., "Prothrombin Deficiency Results in Embryonic and Neonatal Lethality in Mice," *Proceedings of the National Academy of Sciences USA* 95 (1998): 7597–602.

22. Y. Jiang and R. F. Doolittle, "The Evolution of Vertebrate Blood Coagulation as Viewed from a Comparison of Puffer Fish and Sea Squirt Genomes," *Proceedings of the National Academy of Sciences USA* 100 (2003): 7527–32; R. F. Doolittle, Y. Jiang, and J. Nand, "Genomic Evidence for a Simpler Clotting Scheme in Jawless Vertebrates," *Journal of Molecular Evolution* 66 (2008): 185–96; M. B. Ponczek, D. Gailani, and R. F. Doolittle, "Evolution of the Contact Phase of Vertebrate Blood Coagulation," *Journal of Thrombosis and Haemostasis* 6 (2008): 1876–83; R. F. Doolittle, "Step-by-Step Evolution of Vertebrate Blood Coagulation," *Cold Spring Harbor Symposia on Quantitative Biology* 74 (2009): 35–40; "Bioinformatic Characterization of Genes and Proteins Involved in Blood Clotting in Lampreys," *Journal of Molecular Evolution* 81 (2015): 121–30.

23. T. Halkier, *Mechanisms in Blood Coagulation, Fibrinolysis and the Complement System* (Cambridge: Cambridge Univ. Press, 1992), p. 104.

Acknowledgments

1. Lehigh Univ., Biological Sciences, "Department Position on Evolution and 'Intelligent Design,'" https://www.lehigh.edu/~inbios/News/evolution.html.

CREDITS

Figure 1.1: Aaron Bacall, www.CartoonStock.com.

Figure 2.2: From A. Reichenbach and A. Bringmann, "New Functions of Müller Cells," *Glia* 61 (2013): 651–78. Copyright John Wiley & Sons. Reprinted with permission.

Figure 2.3: A. Komeili, "Molecular Mechanisms of Compartmentalization and Biomineralization in Magnetotactic Bacteria," *FEMS Microbiology Review* 36 (2012): 232–55. Permission conveyed through Copyright Clearance Center, Inc.

Figure 2.5: From J. Ruan et al., "Architecture of a Flagellar Apparatus in the Fast-Swimming Magnetotactic Bacterium MO-1," *Proceedings of the National Academy of Sciences USA* 109 (2012): 20643–48. Reprinted with permission of the National Academy of Sciences.

Figure 4.1: David Eppstein, Wikimedia Commons, public domain.

Figure 5.1: Simon Greig, Shutterstock.

Figure 6.1: From P. R. Grant and B. R. Grant, *How and Why Species Multiply: The Radiation of Darwin's Finches* (Princeton, NJ: Princeton University Press, 2008). Republished with permission of Princeton University Press. Permission conveyed through Copyright Clearance Center, Inc.

Figure 6.2: From T. D. Kocher et al., "Similar Morphologies of Cichlid Fish in Lakes Tanganyika and Malawi Are Due to Convergence," *Molecular Phylogenetics and Evolution* 2 (1993): 158–65. Permission conveyed through Copyright Clearance Center, Inc.

Figure 7.2: Liliya Kulianionak, Shutterstock.

Figure 9.1: Ilin Sergey, Shutterstock.

Figure 9.2: Yutanga, iStock.

Figure 9.5: Schab, Shutterstock.

Figure A.1: From D. Voet and J. G. Voet, *Biochemistry*, 2nd ed. (New York: Wiley, 1995). Copyright © 1995 by John Wiley & Sons, Inc. Reprinted with permission.

Figure A.2: Gearstd, Shutterstock.

INDEX

Page numbers of figures appear in italics.